T0064599

ESSAYS FROM THE EDGE OF SCIENCE

ESSAYS FROM THE EDGE OF SCIENCE

Kenneth W. Behrendt

authorHOUSE®

AuthorHouse™
1663 Liberty Drive
Bloomington, IN 47403
www.authorhouse.com
Phone: 1 (800) 839-8640

Published by AuthorHouse 04/28/2015

ISBN: 978-1-5049-0759-0 (sc)
ISBN: 978-1-5049-0757-6 (e)

Table of Contents

Introduction

THIS IS NOT THE BOOK I had intended to be my fifth published work. Rather, I had intended to do a book about an early 18th century craftsman / inventor named Johann Ernst Elias Bessler and the marvelous inventions he had created such as *working* perpetual motion wheels! However, the research I had been conducting into the internal mechanics of his wheels was not yet complete and ready for presentation and several years had already passed since my last work, *The New Science of the UFO*, had appeared. So, in order not to disappoint those who may have been following my writings over the years, I decided to produce the present volume.

In this book the reader will find a somewhat unusual collection of what may be referred to as "New Age" writings on subjects that range from the strange powers of various animal species on and below our Earth's surface to ghosts to religions to UFOs and on to the "big" picture when it comes to understanding the ultimate nature of the cosmos we inhabit. (There's even a final chapter that presents a system for playing and, hopefully, winning state Lotto games! Its inclusion was a last minute decision for reasons that are given at its beginning.)

Much of the material is somewhat old, but has been carefully rewritten, updated, and expanded. Some of it is new and should appeal to the more technically minded reader.

My goal is to impress upon the general reader that current science is still a long way from being "complete" in any true sense of the term. It is, as it has always been, *provisional* in nature and subject to revision whenever new data from experiments and observations is available that contradicts previously accepted theories. Change, however, does not come easily to any human endeavor, especially science.

For example, Lord Kelvin, the greatest scientist of the 19th century, declared in 1899 that the science of his day was complete and that nothing new would be added to it other than more precise measurements due to improvements in instrumentation. Little could he know then what massive changes would come to the world of physics when the works of such notables as Niels Bohr and Albert Einstein appeared only a few years later. Their research was eventually accepted as valid, but only after much debate and opposition.

No doubt the material presented in this volume will also be dismissed by the current world of science as just so much fantasy. However, I remain highly confident that most of the material contained herein, particularly that which is more technical in nature, will be found in our university physics textbooks (or their digital equivalents) by the end of this 21st century and, perhaps, much sooner in certain cases. Thus, the reader should consider this volume as a sort of crystal ball that will give him or her an advanced look into what future science will look like.

As with all of my works, I advise the reader to start at the beginning and work his way through the chapters of the book in the order in which they are presented. This is because I always try to arrange material so that it builds upon itself. If one reads the chapters out of their intended order, he may encounter material that involves concepts developed in earlier chapters with which he is not yet familiar and then confusion can result.

Hopefully, the reader will, by the time he completes this volume, be impressed that we live in a world wherein there is real magic occurring which is still just beyond the edge of currently known and accepted science. Soon the physics of that magic will be understood completely and as a result we will see major paradigm shifts taking place in science and philosophy which will then quickly affect, most likely for the better, practically every aspect of human life on our planet and its relationship to the rest of the cosmos.

Chapter 1

Animal Intelligence and Telepathy

I WAS A CHILD WHEN I had my first experience with an "intelligent" animal. It happened on one of several weekend sleepover visits that my parents would take me along on when they visited an aunt of mine that lived in a neighboring state. She owned a most remarkable male collie that was named "Duke" or, more affectionately, was just called "Dukie". I always looked forward to those visits to my aunt's home because I loved playing with her dog. The reason was that he had a most unusual ability which I have never seen any other dogs display. Dukie apparently understood English to a degree which almost made him seem human!

For example, if I told Dukie to follow me, he would and remain within a few feet of me at all times. If I told him to go up or down the stairs by himself, he would. And, if he was anywhere in the house and I called him, he would immediately be at my side. His ability to understand English became most apparent when he was instructed to fetch an item. I could tell him to bring me a ball, slipper, or newspaper and he would. If he brought the wrong item and I said "No, that's not right", he would immediately return it to its original location and continue to select other items until he got it right.

I remember one incident, in particular, when I noticed that he seemed to have a limitation in his abilities. While he could discern differences between the sizes and shapes of objects, he could not tell the difference between their colors. Once, there were two sponge rubber balls that I had brought with me. One was red and the other blue in color. Dukie seemed unable to tell the difference between them when he was asked to fetch one of a particular color. It was only in later life that I finally learned why. The reason was that dogs, for some strange reason, lack the retinal cells in their eyes that would allow them to perceive colors. It is thought that this may stem from their evolution as nocturnal hunters whose vision must be more specialized for seeing in low light levels rather than for distinguishing between colors which, apparently, is easier to do during the daylight hours.

I believe that much of Dukie's remarkable, almost human abilities, were due to the fact that my aunt raised him from a puppy and that he lived, unrestrained, in the house with her. Each night he would sleep at the foot of her bed. He had little interaction with other dogs and probably considered himself to be human!

Dukie had a long life and provided my dear aunt with many years of companionship. Eventually, as he grew feeble and incapacitated from age, it became necessary to have him "put to sleep". He died in my aunt's arms while organ music was played in the background and his passing was something that deeply affected all of us for many months afterwards.

As I grew up I heard of other collies which also displayed unusual intelligence. Of course, as a child I always watched the TV show "Lassie" whose canine star reminded me a lot of Dukie. I also remember that during the late 1980's there was yet another collie that was featured in a television commercial for some fast food restaurant chain. This dog had the quite unusual ability to mimic human speech with enough clarity to be understood!

For example, in the commercial his owner proclaims to someone that nothing is impossible because he had just taught his dog to say "I love you"! The collie then says those words and they are clearly recognizable. For a dog to be able to mimic human speech like that requires a high degree of intelligence. He must be able to recognize when the sounds he makes differ from those of his trainer and be able to modify them to get them as close to those of the trainer as possible. This task is akin to a human trying to repeat words in a foreign language without actually understanding the meaning of the words.

I have, over the years, heard or read of many other interesting cases of canine intelligence. For example, the recent tragic tsunami that claimed hundreds of thousands of lives in the countries surrounding the Indian Ocean produced many interesting stories of animal intelligence and, perhaps, precognition. One involved a small dog who, on the morning of the disaster, began barking excitedly. This dog began nipping at the feet of a young boy who was still asleep in an effort to awaken him. Once awake, the dog literally tugged the boy out of bed and began dragging him toward the door of their dwelling. Once outside, the dog pulled the boy in the direction of high ground until both were tens of feet above their previous elevation. Only a minute of so after this, the house that the boy and dog had occupied was swept away by a rapidly moving twenty foot high wall of water!

It is interesting to note that, despite the enormous loss of human life, very few animals were killed by the tsunami. Somehow they sensed the approaching water and took immediate evasive action. Elephants kept for labor were breaking their chains and moving inland as fast as they could run. Rather than telepathy or remote viewing, these cases of animal precognition are, most likely, due to their ability to sense the low frequency ground waves that reach their locations from an undersea earthquake that creates a tsunami. Thus, they actually have the most advanced warning possible when such a disaster is about to occur.

There are, of course, many tales of animal telepathy that I have picked up over the years. One involves the actions of wolves when traveling in packs.

It has been noted through observation on many occasions that the she-wolf apparently has strong telepathic abilities. If one of her cubs should break away from a traveling pack and be too far off for the mother to quickly reach it, the mother she-wolf will stand rock still and stare intently at the distant cub. The cub will invariably also freeze in its tracks and immediately return to the pack. This strange behavior has also been noted for several other canine species. In such cases, we have evidence that telepathic commands can be transmitted over considerable distances and that the process involves the use of the sender's *and* receiver's *eyes*.

In past writings, I have hypothesized that the transmission phase of mental telepathic communication involves the rapid translation of mental images in a sender's mind into a sequence of extreme low frequency or "ELF" electromagnetic radiation "trains" that are generated within and then emitted simultaneously from *matched* pairs of small rectus muscles that are attached to the sender's *two* eyeballs. (Note that a single ELF train consists of a brief series or "burst" of connected *individual* electromagnetic oscillations that has a fixed frequency and contains a finite number of oscillations.)

During the telepathic transmission of a mental image, *each* edge of the image's shape is rapidly and unconsciously scanned and, as this is happening, neural impulses are sent down the sender's optic nerves to matched pairs of rectus muscles in his eye sockets which *try* to cause the rotations of his two eyeballs to take place such that their two imaginary projected optical axes will then trace out that particular edge's shape. For example, if the optical axis of each of a sender's eyeballs tries to trace out, say, the top *horizontal* edge of the shape of a mental image in his mind by moving from the right to left end of that edge, the single rectus muscle attached to the *left* side of *each* eyeball will *try* to contract a bit so that the two eyeballs will together undergo a gross rotation to the left.

I wrote above that the two rectus muscles involved "try", via contraction, to cause a gross rotation of the sender's eyeballs to the left, but, fortunately, they do not succeed because as soon as they begin to produce such rotations, there is an automatic feedback or reflex mechanism activated (which involves the visual cortex of the sender's brain) that almost instantly sends nerve impulses to the matched pair of *opposing* rectus muscles that are attached to the *right* side of each of the sender's two eyeballs. These neural impulses then cause these two opposing rectus muscles to contract just enough so that the counter torques they apply to the two eyeballs actually prevents them from undergoing any gross rotational motions at all.

Once this blocking action has occurred, *another* section of the top horizontal edge of the sender's mental image a tiny bit to the left of the first section is considered and the original matched pair of rectus muscles in his two eye sockets then once again try to rotate his two eyeballs to the left so that their imaginary projected optical axes would then trace out that new section of the edge of the mental image. But, again, the opposing rectus muscles are reflexively activated in order to frustrate this second attempt. This process is rapidly repeated as the entire top horizontal edge of the mental image is continually and progressively "scanned" in the sender's mind during the telepathic transmission process. This scanning process requires no conscious effort on the part of the sender.

Eventually (in just a matter of milliseconds!), the entire length of the mental image's top horizontal edge will be scanned. While this is happening, all *four* of the little muscles in the *two* opposed matched pairs of rectus muscles that are attached to opposite sides of the sender's two eyeballs (i.e., two muscles of one of the matched pairs will be attached to the *left* sides of the sender's two eyeballs and two muscles of the other opposing matched pair will be attached to the *right* sides of his two eyeballs) will be in a state of continuous *suppressed* tremor due to the counteracting contractive forces that they are alternately and rapidly applying to each other. It is these very, very minute tremors that take place in the rectus muscles involved as the shape of a mental image is telepathically transmitted that are, I believe, ultimately responsible for all of the ELF trains emitted from a sender's two eye sockets.

As soon as *one* of the edges of a mental image has finally been completely scanned, the rectus muscles involved will all cease to tremor and none will emit any more ELF trains unless, of course, it is recruited to scan another edge in the mental image's shape. At all times during this scanning process the sender remains consciously unaware of the rapid, but suppressed, tremors taking place in his two eyes' rectus muscles and these minute tremors have no effect on his vision.

So, we see from this that, as each edge (whether horizontal, vertical, or diagonal) of a sender's mental image's shape is scanned during the telepathic transmission process, there are actually four ELF trains being emitted from the sender's two eye sockets. Two of the ELF trains are produced by the matched pair of rectus muscles as they repeatedly try to grossly rotate the sender's eyeballs so that their optical axes will smoothly trace out the edge of the shape of the mental image while the other two ELF trains are produced by the matched pair of *opposing* rectus muscles as their actions repeatedly block the eyeball rotations that would be caused by the first matched pair of rectus muscles if they were unopposed.

As all of the edges of the shape of a mental image are rapidly scanned during the telepathic transmission process, the resulting sequence of various ELF trains generated will radiate out of the interiors of the rectus muscles involved and then immediately enter the fluid interiors of the sender's two eyeballs which are attached to these rectus muscles. There the ELF trains are "condensed" or gathered together by the fluids and finally issue from the sender's two eye sockets so as to move along at light velocity within a narrow divergent beam toward the general direction that the sender faces.

Each eye socket actually emits its own single beam, but for any animal with two eyes and stereoscopic vision, the sources of the two beams generated are so close together that their overlap effectively forms them into a single invisible beam. (Indeed, it should eventually be possible to detect and even make visible these beams through the use of special equipment. Once that is done, it will then be a relatively easy matter to produce such a beam artificially so that any kind of image can be transmitted with it. In such a case the image being transmitted will not be one created by a living brain, but rather by a computer program!)

In the receiver's eye sockets, the exact opposite process will take place as occurred in the sender's eye sockets during transmission of the shape of the mental image in the sender's mind.

An incoming matched pair of ELF trains from the sender's *two* eye sockets will, via the fluids in the receiver's eyeballs, be directed into the matched pair of rectus muscles attached to the receiver's two eyes that *anatomically* corresponds to the *same* matched pair of rectus muscles attached to the sender's eyes that originally generated that particular matched pair of ELF trains. The absorption of these two trains' electromagnetic energy will then stimulate the two rectus muscles in the receiver's matched pair (consisting of one rectus muscle attached to each of his two eyeballs) to begin to contract so as to rotate his two eyeballs in a certain direction.

But, almost simultaneously, the receiver's eye sockets are also picking up the incoming matched pair of ELF trains that were generated by the *opposing* matched set of rectus muscles attached to the sender's two eyeballs. This second incoming matched pair of ELF trains is then directed into the *opposing* matched pair of rectus muscles attached to the *receiver's* two eyeballs and stimulates them to begin to contract so as to try to cause the receiver's eyeballs in rotate in the opposite direction to that which would be caused by the first matched set of incoming ELF trains *if* their action was unopposed.

Thus, the eyeball rotation producing action of a particular stimulated matched pair of rectus muscles attached to the receiver's two eyeballs is almost immediately counteracted by the stimulated action of the *opposing* matched pair of rectus muscles attached to his two eyeballs. This results in the same minute tremors taking place in the receiver's rectus muscles as took place in the sender's *same* rectus muscles during the telepathic transmission of the shape of the mental image in the sender's mind. These suppressed tremors in the receiver's rectus muscles then result in the generation of neural impulses that travel up the receiver's two optic nerves and into the visual cortex of his brain to create an image that he then becomes consciously aware of. (Technically speaking, the suppressed tremors in the receiver's rectus muscles should also transmit the same mental image back to the sender. However, this feedback transmission is probably so weak that it can not interfere with the process of telepathic communication.)

Since only milliseconds of time are required for the scanning of a mental image in the sender's mind, its transmission through space to the receiver's eye sockets, and its conversion back into a conscious image in the receiver's mind, the result is that the mental images in the sender's mind are almost immediately reconstructed in the mind of the receiver as mental telepathic communication takes place.

So far we have only covered telepathic communication which involves the transmission of mental *images* from one mind to another. But, humans generally communicate with each other by sending sound wave patterns known as "words" back a forth between themselves and not by sending pictures back and forth. Interestingly enough, verbal communication also seems to be possible using the telepathic processes described above.

There have been a limited number of highly reliable UFO abduction cases in which the human captive later, during time regression hypnosis, claimed that he was able to verbally speak to and understand the verbal responses of the extraterrestrial beings with whom he was interacting. Obviously, such beings would not be fluently speaking any of Earth's many languages so telepathy must have played a major role in these very rare communications. Let us now briefly consider how such verbal telepathic communication might work.

When a human being "subvocalizes" or *thinks* of the sounds of a word in his mind silently, he usually does not also "see" a mental image associated with that word "in his mind's eye" unless he tries to do so. However, just because there is no conscious mental image formed as he subvocalizes sounds does not mean that such an image is not formed in his *sub*conscious mind. That subconscious image, although far weaker than one held in his conscious mind, is also immediately scanned and causes the appropriate rectus muscles in his eyes to begin weakly emitting the ELF trains associated with that image.

If another nearby person's eye sockets happen to receive the resulting low power transmission of the sender's weak subconscious image that was associated with the sounds of a certain word he was subvocalizing, then that weak image will not be strong enough to cause a copy of itself to form in the receiver's conscious mind. However, that weak image will still be able to form in the receiver's subconscious mind.

Apparently, this weak image in the receiver's subconscious mind can then immediately trigger the recall of the particular sounds that the *receiver* associates with the image to enter his conscious mind and he will then "hear" those sounds which will, to him, appear to be "in his head". If, by chance, the sender and receiver happen to have attached the same image to the same subvocalized sounds, then the receiver will actually "hear" the same sounds that are being subvocalized in the sender's conscious mind.

One fascinating aspect of such verbal telepathic communication is that it actually allows two sentient beings to effectively communicate with each other even though they do not speak the *same* verbal language and this rationalizes how meaningful two way verbal communication can take place between a human being and an extraterrestrial being.

For example, an alien being might subvocalize or even audibly and very softly speak his word for "table" to an abductee who he wants to lie on a table inside of his craft so that some sort of medical examination can be performed on the human. The weak subconscious image that the extraterrestrial's mind associates with an actual table is then transmitted to the human's subconscious mind and immediately triggers the recall of the sounds for the word "table" in the human's own language and he actually "hears" these sounds in his mind. Whenever the human verbally responds to his abductor, the human simultaneously transmits the weak subconscious images associated with his words to the alien's subconscious mind and he immediately "hears" the sounds associated with these images in his own alien language.

This process for verbal telepathic communication will, of course, work for entire sentences and serves as a form of universal translation that allows any two sentient beings to communicate with each other to some degree. How effective that communication will be depends, most likely, in how similar the body types of the two beings are. If they are both humanoid in shape and come from similar planets, then when each subvocalizes or actually vocalizes the sounds for his word that describes a "house", then the weak subconscious mental images associated with those sounds will be very similar.

Whether for images or sounds, the telepathic communication process, as in the case of the transmission and reception of ordinary radio waves, becomes far more efficient when the two minds involved are "attuned" to each other. The proper tuning is most likely to exist between close relatives within a *single* species and accounts for the she-wolf's ability to best be able to call only her own cub, but not another wolf's cub. Only very rarely will such telepathic attunement exist, by chance, between genetically unrelated members of the same species or between members of different species and this helps prevent a predator from using telepathy to lure prey of a different species into a situation where it could be attacked.

Even animals with brains much smaller than canines can demonstrate astonishing levels of intelligence. Recently, I saw a televised story that came out of England. It concerned research done with common crows that indicated that they are the "Einsteins" of the bird world.

In a college anthropology course years ago, I was told that only humans and a few of the other higher primates had enough intelligence to use tools to accomplish various difficult tasks. As a demonstration of this, our class saw a film showing how a chimpanzee in a cage eventually figured out how to use a stick to extend his reach and then pull a banana outside of the cage toward him which he ordinarily would not have been able to reach. However, the above mentioned story about crow intelligence was far more impressive.

In the news story, a video clip of a crow was shown. He was in a cage that had a variety of small items scattered about its floor. In the center of the cage was a clear plastic cylinder, one end of which was firmly attached to the floor of the cage. The plastic cylinder stood upright and at the bottom of it was placed a tiny basket containing a tasty nut that the hungry crow wanted to consume. The basket had a little handle attached to it and the crow made several futile attempts to insert his beak through the open end of the plastic cylinder in an effort to bite the handle and then pull the basket and its contents up and out of the cylinder. However, the diameter of the cylinder had been carefully sized so that it was physically impossible for the crow's beak to reach the basket's handle no matter how hard he tried.

9

After a few attempts the crow stopped sticking his beak in the cylinder and began looking around at the small objects scattered on the floor of his cage. He eventually found a small length of wire and picked in up in his beak. He then inserted the other end of the wire into a small hole in one of the other objects in the cage and bent that end of the wire until it was shaped like a small hook. Next, he returned to the plastic cylinder in the center of the cage and, using his beak to hold the still straight end of the piece of wire, he inserted the hooked end of the wire down into the cylinder. The hook was used to snare the handle on the basket and then lift the basket up and out of the cylinder! He then immediately consumed the nut as a reward for his effort.

When I saw this little demonstration, I was totally dumbfounded. The speed with which the bird recovered the nut indicated a very high degree of intelligence in terms of judging distances, the properties of materials, and the logical steps needed to accomplish a complex task.

With such obvious intelligence, one wonders if birds, especially crows, can also display telepathic abilities. Interestingly, I also encountered one story presented on another television show devoted to strange phenomenon that indicated crows, and perhaps most birds, do have very powerful telepathic abilities.

The story involved a woman who described herself as a "pet detective". She would help people in her neighborhood with various problems related to their pets. However, she did not work alone and was assisted by her pet parrot that just happened to be highly intelligent and telepathic!

To prove the psychic powers of her parrot, the lady demonstrated how the bird could literally read her mind. She was told to concentrate on one of her fingers by the people videotaping her and the parrot would then pick out which one of her five extended fingers she was concentrating on. The bird was able to do this correctly 100% of the time. At no time when her fingers were outstretched did they shake or twitch in the slightest, so the possibility that the parrot was only "body reading" the woman was eliminated.

One day, another lady in the neighborhood came to the woman pet detective and was in a desperate state. Her pet cat had been missing for several days and she was hoping that the pet detective could locate the lost animal. The client was told not to worry about the cat which the detective would try to locate as quickly as possible and return to her.

The pet detective then telepathically told her parrot of the situation and the parrot supposedly telepathically responded by saying he would relay the request to find the cat to a group of crows that inhabited some nearby trees and that they, in turn, would pass it along to all of the other crows in the neighborhood. Thus, in a short time all of the crows in the area would be on the lookout for the missing cat and, once it was located, they would help lead it back to its home. All of this relaying of the request between the various birds was also done entirely by telepathy.

About two days after the request was made, the pet detective received a telephone call from the client. She was now overjoyed that her cat was home safe and sound, but the cat's sudden arrival had been a little strange. Its owner had heard the small flap used by the animal at the bottom of her kitchen door open the night that the cat returned and, after welcoming the cat back with a bowl of warm milk, she decided to look around her yard. In the yard there was a tree and perched in it were several crows! Apparently, they had found her cat and managed via telepathy to lead it back to its owner.

Next, let us turn our attention to animals larger than either birds or humans and see what strange intellectual and telepathic powers they can possess.

Another psychic lady whose story appeared several years ago on a strange phenomena television show claimed to also be able to establish telepathic communication with animals. Her specialty, however, was animals such as horses and various creatures kept in zoos.

One day, this psychic lady received a call from the operators of a local zoo. They were desperate for a solution to a mysterious "illness" that had befallen the small group of elephants at the zoo. For almost a month the elephants had barely eaten anything and their health was steadily declining. All of the medical tests given these poor creatures were unable to uncover the nature of the malady. Someone had suggested that the animals had a "psychological" problem of some sort, but that person did not know of any local animal psychologists that could be called into the case. Finally, it was decided to try the psychic because she was known to have had several successes in dealing with animal problems because of her alleged psychic powers. Fortunately, for the zoo keepers, she agreed to try to help the small herd of elephants.

She arrived at the zoo and, with a keeper at her side, moved in among the elephants that were rather wary of this stranger. However, after a few minutes, they adjusted to her presence and began to approach her. Some of the elephants were very weak from not having eaten properly in weeks.

The psychic then walked up to one of the smaller female elephants and began stroking her trunk while maintaining close eye contact with her. After about a minute of this, the lady psychic turned to the keeper and announced that all of the elephants were not eating because they were grieving! Apparently, about a month earlier, one member of the small herd had died suddenly and the zoo officials had its body promptly removed and buried. The psychic claimed that the female elephant had told her, telepathically, that all of the elephants were very sad and depressed that they had not had a chance to say farewell to their deceased friend!

At first hearing of all this, the zoo officials thought the whole idea was ridiculous. However, one member of the herd had died a month earlier and the lady psychic had not known about it in advance of her session with the elephants.

Over the course of the next few days, the remains of the buried elephant were exhumed and one of its thigh bones was retrieved and brought back to the zoo. The lady psychic then brought the bone in among the elephants and its mere presence seemed to energize them. One by one they picked up the bone and passed it back and forth among themselves while making low pitched mournful sounds. When they were done, the bone was placed in a corner of their pen where they could see and touch it whenever they wanted.

About a week later, the psychic received a phone call from the zoo's director. He was very happy to report that the mysterious malady that had stricken the elephants was gone and they were rapidly regaining their health and eating normally again.

There is one more interesting story related to elephant intelligence that I remember reading about a few years ago. This one took place on a banana plantation on the island of Sri Lanka which is located near the southern tip of India.

One day workers on the plantation approached the owner to notify him that, for several nights in a row, someone had been trespassing on his plantation and stealing large quantities of bananas. The owner decided to finally put a stop to the theft and hired armed guards to patrol the perimeters of the large piece of land. However, this proved useless and the thefts continued every night without fail. What made the situation even more mysterious was that none of the guards heard or saw anyone during their nightly patrols of the property's borders.

The owner grew more desperate to put a stop to the nightly thefts of his valuable crop and decided to place armed guards into the banana fields themselves. This time, the guards were instructed to hide themselves so that they might catch the thieves in the act.

The plan was successful on the very first night that it was tried. Early in the morning, the guards in one of the fields began to hear the sounds of banana plants being disturbed and they slowly moved toward the source of the sounds. What they found amazed and stunned all of them and the owner was quickly awakened so he could confront the thieves personally.

Huddled together and frightened by the sudden appearance of gun toting, flashlight waving humans was a small group of elephants! They belonged to a local man who hired them out to perform various manual labor chores like moving logs or pulling stuck vehicles out of the mud.

Apparently, the creatures had not been properly fed and decided to solve the problem themselves. Every night for several weeks they had been letting themselves out of the enclosure in which they were kept. They would then walk, single file, down to a small nearby stream. There they would take a refreshing drink and then carefully stuff mud from the stream's banks into the bells they wore around their necks so that these could no longer sound their presence and movement. Finally, they would slowly make their way to the plantation for their nightly snack.

The story has a happy ending. The elephants were not harmed, their owner paid for the damage they had done, and their meal sizes were substantially increased. Also, greater attention was paid to the security of their enclosure to prevent them from making any more nocturnal visits to the neighboring plantation.

No chapter on animal intelligence and telepathy would be complete without some mention of horses. Of course, most older Americans are familiar with Roy Rodger's famous horse "Trigger" who could be summoned with a whistle by Rodger when he needed him. This horse, like my aunt's collie, could also respond to verbal commands and seemed almost human at times.

I remember seeing a segment of a television documentary a few years ago about a woman who kept an adult horse as a pet. She had turned a large ground floor room in her home into a stable for the animal and had arranged its construction so that a window over her kitchen sink opened into it.

In the segment, as she began cleaning the dishes after her family's dinner, the window over the sink suddenly opened up and the horse stuck his head into the kitchen. He wore a large floppy hat and took a scrubber that was attached to a stick from the lady. He held this in his teeth and proceeded to scrub out one of the soaped up pots in the sink! Apparently, he considered himself one of the members of the family and wanted to pitch in and help with the dishes after dinner.

Outside the house, the horse could run free over the entire property, but would immediately come running back if the lady called to him. Toward the end of the segment, an interesting incident happened which particularly caught my attention.

The lady approached the horse that she had just summoned and began petting his nose. He liked this very much. Then, the lady wrapped her arms around his neck and proceeded to hug him. He responded by using his chin on the lady's back to gently pull her toward him. It seemed to me that he wanted to return her embrace, but realized that he could not use his front legs to do so. Instead he used his chin and carefully adjusted the force so as to not injure his owner. To act in such a manner demonstrated to me just how highly intelligent and, indeed, loving this creature was.

As a final story about horse intelligence, I want to mention the case of "Clever Hans". He was owned by a retired German school teacher named Wilhelm von Osten who, in the 1880's, was convinced that he had been able to teach the horse to perform mathematical calculations.

During public demonstrations, Hans the horse could add and subtract two digit numbers and even calculate the squares of numbers. Von Osten would verbally ask Hans to solve a math problem and the horse would respond by tapping out the number in the answer by using his front hoof. He rarely made a mistake in these demonstrations. Hans could also provide word answers to questions by laboriously tapping out the letters in words of the answer. For example, for the letter "A" he would tap once. For the letter "B" he would tap twice. For the letter

"C" he would tap three times and so on for the remaining letters of the alphabet.

As Hans' fame slowly spread throughout Europe, it was not long before he began to attract the attention of the scientific community. At first, only individual scientists, mainly early psychologists, examined the horse and most left convinced that he really could perform simple mathematical computations. Much debate began to swirl around the issue of whether the horse was actually performing math calculations or whether he was somehow being cued to provide the right answers by his owner so that a fraudulent profit could be made off of his exhibitions.

Finally, a commission of distinguished scientists was assembled whose goal was to determine, once and for all, if Hans was really as clever as he was promoted to be. The commission set to work testing the horse and it was immediately realized that he provided correct answers even when his owner, von Osten, was not present! This eliminated the owner as a source of any fraud. However, then, quite by chance, an interesting observation was made.

It became apparent that Hans could not give accurate answers to math problems when nobody present in the room he was in knew the answer! In one test, the numbers from 0 to 9 were written on large cards and positioned so that only Hans could see them on the ground. The witnesses then asked him to solve a simple addition problem by tapping on the card with the answer. In every case, he performed no better on this test then random chance. However, when any witness in the room was positioned so that he could also see the cards, Hans was again able to select the right answer with 100% accuracy!

Tests like these (and there were many others) proved beyond a shadow of a doubt that Hans' math ability was very poor, but his ability to "body read" the scientists working with him was excellent. Apparently, after being asked to solve a math problem, he would slowly begin tapping out an answer and would carefully observe the subtle, unconscious reactions of the person who had asked the question. As he approached the right answer, the human might suddenly stiffen his posture a bit or stop blinking and then Hans would know that he had the right answer and stop tapping. The horse did this neat little trick so smoothly that he managed to fool some of the top people in the field of psychology of his day.

However, I do vaguely remember reading about an interesting story connected with the testing of Hans, the veracity of which I can not verify. Apparently, during his early testing by individual scientists, one scientist was allowed to be in a tent with him alone and ask him questions which he would then answer by slowly tapping out the letters of the alphabet in reply. Supposedly, he was asked the question "Why don't you speak?". His replay was, "Can not talk, throat no good"! If this is a true story, then Hans was very intelligent indeed. Perhaps Hans had tried to verbally respond to the questions posed to him, but soon realized that he could not articulate human sounds with his horse's throat. He would, thus, have been aware that he was different from the humans who questioned him.

Of course, the legends of humanity contain many stories of animals that could speak and / or communicate telepathically with human beings. During his travels around the ancient world in an effort to promote early Christianity, St. Paul supposedly encountered a talking lion somewhere in what is now modern Turkey that requested to become a Christian. After Paul had him repent of his sins, he was then baptized into the Christian faith! The legends that Bram Stoker drew upon to create his famous character of Count Dracula claimed that vampires could telepathically communicate with and control various animals such as rats, bats, and wolves.

Since I am a firm believer in the reality of telepathic communication (I describe a personal experience in my first book, *The Physics of the Paranormal*), I believe that one day humans will be able to routinely communicate with the animal life with which we share this planet. Such communication is, as this chapter demonstrates, already occasionally taking place reliably. However, it may even someday soon be possible for anybody, via technology, to effectively engage in two way, mind to mind telepathic communication with any animal whose intelligence is equal to or greater than, perhaps, that of a bird or a mouse.

Since all of the evidence I have managed to gather indicates that the process of mental telepathy uses the emission of trains of extreme low frequency or ELF electromagnetic radiation that are transmitted back and forth between the eyeball rectus muscles of two separated organisms, we should be able to eventually determine which sets of trains correspond to particular mental images and then use some sort of detector coupled to a high speed computer to intercept and then rapidly translate these ELF trains back into the visual images inside the mind of the brain that generated them along with any sounds associated with those images.

Incredibly, it might eventually even be possible to construct a receiving / transmitting device that would pick up *any* target animal's telepathic emission and then instantly convert its ELF trains into the types suitable for reception by a particular human's rectus muscles. Once received, these "translated" trains would then stimulate the person's rectus muscles as described earlier and the resulting neural impulses would travel up his optic nerves and into his brain's visual cortex where they would form strong conscious mental pictures and along with any sounds associated with these images.

Using this device in reverse would then allow its human operator to instantly send a meaningful message to a target animal in a form which that particular species of animal could receive and understand. Perhaps the device could even be designed to pick up and analyze any incoming telepathic ELF trains from an animal and then use this information to immediately determine what species of animal was to be the recipient of the human sender's telepathic response. The device would thus automatically adjust itself to facilitate telepathic communication between a human mind and that of a particular species of animal.

With the miniaturization possible in electronic equipment, this animal / human telepathic communicator might be made small enough so that it could be housed in an ordinary pair of eyeglasses! The human using it would need only stare at any particular animal and would almost immediately be aware of what was going on in that animal's mind. A scientist working in the field with animals would realize why they were behaving as they did. He would know their desires and fears, their impressions of each other and their environment, and what their reactions were to the presence of humans. A veterinary doctor could query his animal patient about the nature of a disease or injury that the animal acquired and, thereby, obtain a more accurate diagnosis for a more effective treatment.

Imagine the benefits of being able to safely travel through any forest or jungle knowing that one could immediately communicate to the surrounding wildlife that one meant them no harm. One might even be able to enlist the aid of the surrounding animal population in finding food or water or one's bearings if lost. A person who was injured might be able to send an animal messenger such as a bird to the nearest human outpost to seek medical assistance. The animal could then telepathically give rescue personnel there a precise location for the injured person, his condition, and even lead them back to the person.

Once telepathic communication with the other members of the animal kingdom is routinely established, humans will, I believe, have a new and far more responsible attitude toward the environment of planet Earth. We will realize that, while our actions may not harm human health, they might endanger the health of the intelligent creatures that live in the wild. To meet our responsibility to our fellow lifeforms, we can expect even greater concern for environment protection. All polluting industries will, eventually, have to be replaced with environmentally friendly ones. Gone will be polluting factories and automobiles. Power generation via combustion and risky nuclear energy will have to go. All products for human consumption will have to come in non-toxic biodegradable containers. And the list of changes necessary only begins here!

Meeting these demands will be an awesome challenge. The consumption of meat will have to give way to a totally vegetarian diet that is free of the contaminants now found in meat and its byproducts. All synthetic materials will have to be replaced with natural products whenever possible. We will all, as a result, have simpler and healthier lives with far less disease and disability. The changes will be hard to make, but, I suspect, once they are, humanity will never again return to our present mode of living which we consider to be "modern".

Will we make these changes? When will they occur? These are very hard questions to answer from the perspective of the early 21st century. However, I think it not unreasonable to expect them sometime during the next several centuries. Indeed, I would be surprised if most of the changes I envision above were not in place by the year 2100 AD.

Chapter 2

Our Forbidding Ocean World

I REMEMBER GOING TO THE THEATER one day many years ago and seeing the Walt Disney movie titled "20,000 Leagues Under the Sea" which was based upon the Jules Verne science fiction novel from the 19th century. It is an absorbing tale that starts with a ship that is sunk after being rammed by a mysterious sea monster. Some of the survivors from the wreck are eventually rescued by the "monster" which turns out to be a fabulous submarine built by a genius inventor.

The inventor is a man named Captain Nemo and we learn that he constructed his submarine, the "Nautilus", as a weapon to wage his one man battle against the warring land governments of Earth. As the people he rescues at sea soon discover, his underwater ship is actually nuclear powered (a remarkable prediction to have been made by Verne in the 19th century!). The Nautilus can travel so fast through the water that, with its heavily constructed metal hull and pointed bow, it is able to ram and tear the hulls of wooden vessels apart causing them to quickly sink along with any of the weapons and ammunition that they might be carrying that enable the world's land governments to continue their warfare. Of course, in the process, most of the crews and passengers of these ships are also killed even though they might not have any responsibility for the military cargos being carried.

The Disney film, like the original novel, takes us inside the underwater realm of Nemo and we see what it might be like to actually live and work underwater. All of the comforts needed to survive in the dark depths of the world's oceans are at Nemo's disposal and he is, in fact, a kind of ruler of a kingdom that covers about 70% of the Earth's surface. He has air, drinking water, food and sophisticated electrical weapons that allow him to protect himself from the various gigantic creatures that also share his aquatic realm. Nemo also has access to all of the treasures ever lost at sea during shipwrecks. He is not completely independent of land, however. After completing a journey of 20,000 leagues while traveling underwater (which is 60,000 nautical miles or 69,047 statute miles and equal to a distance that is about 2.778 times the circumference of the Earth at its equator!), the Nautilus must be taken back to its secret volcanic atoll base in the South Pacific Ocean where it was constructed for various maintenance repairs and refueling.

Much of the story revolves around the attempts by the "guests" aboard Nemo's submarine to escape their captivity and return to their own countries. I won't spoil the ending for anybody who has not seen the Disney film or read the original book by revealing all of the details here. Let it suffice here for me to just add that the character of Captain Nemo is, perhaps, one of the most fascinating in literature. He is simultaneously both hero and villain. A genius in pursuit of a peaceful world who, like the modern terrorist, has along the way somehow lost concern for the innocent lives he is taking. As a science fiction story, "20,000 Leagues Under the Sea" is probably in the top ten ever written and well worth reading by anybody interested in science or technology. Perhaps of all the science fiction works ever created, it is truly remarkable for the detail of its accurate predictions about the future of such things as submarines and nuclear power.

Many people have probably read underwater adventure stories like the above and scoffed at the idea that Earth's oceans could harbor creatures the size of the ones Captain Nemo had to combat. They may agree that there are large whales out there, but these rarely exceed a length of about 60 feet and really are not that terrifying. They usually pose no real danger to larger surface vessels or submarines.

The sperm whale is the largest of the 32 species of whale found in the genus *Physeter* and males can reach lengths of 60 feet with body weights of about 50 tons. Their common name derives from a waxy substance called "spermaceti oil" that is produced by a large organ in the head of each whale. On average each adult whale carries about three *tons* of this material in its head! This substance was once very valuable and in the 19th and early 20th centuries found use in various cosmetic and pharmaceutical products and even candles. The material was extracted from slaughtered whales and could be processed to produce a light, white crystalline substance that was easily powdered for commercial use. The outlawing of whaling stopped the supply of this material which then caused many developed nations to hoard supplies of it. The reason was that it is also an excellent high temperature lubricant which finds use inside of nuclear reactors. Fortunately, there are now synthetic alternatives to it which perform as well or better.

Sperm whales are extremely intelligent creatures with brains that can weigh up to 20 pounds which is far larger than that of a human being. They also live about as long as a human or about 70 years. The whaling industry almost hunted them to extinction, but the banning of whaling is allowing their population to slowly increase. Currently, there are only about 200,000 of them worldwide.

Some recent oceanographic discoveries have revealed a rather perplexing behavior on the part of whales.

There is now well established evidence that the sperm whale or *Physeter macrocephalus* can dive to depths previously unsuspected. In fact, deep water submersible craft capable of operating at ocean depths down to almost three miles (or 15,840 feet) have recorded some of their whale "songs" at depths deeper than 10,000 feet! That's almost two miles straight down and, starting at that depth, one would have to rise through a distance equal to about *eight* Empire State Buildings stacked on top of each other in order to reach the surface of the ocean. The water pressure at that depth is a crushing 4,430 pounds or about 2.2 *tons* per square inch!

In order to survive the enormous crushing pressures at these depths, a whale must have to do something rather unique.

Just before beginning a descent from the ocean's surface, a whale probably hyperventilates through the "blow hole" located at the top of the front portion of his head (the orifice is actually a 20 inch long, S-shaped hole on the left side of the head). This action then results in a build up of a large supply of oxygen containing oxyhemoglobin molecules in his red blood cells as well as saturating his blood serum and body tissues with dissolved oxygen molecules. He then begins his dive and, as the external air pressure begins to get uncomfortable for him at a depth of, perhaps, a hundred feet or so, he will slowly begin releasing the air from his huge lungs and allow these to become flooded with sea water (some very small amount of highly compressed air is, however, retained otherwise it could not be slowly released to make internal membranes vibrate and thereby produce a "song" at great depths). Simultaneously, he also allows his gastrointestinal tract to completely fill with sea water by simply opening his mouth a bit.

As his dive continues, the water that flooded out his lungs and gastrointestinal tracts serves to resist the slowly increasing pressure of the sea water that surrounds his body. Thus, there will actually be no danger of his lungs or gastrointestinal tracts rupturing and then collapsing from the enormous external water pressure acting on the outer surface of his body. In fact, by allowing the sea water in his internal tracts to have continuous contact with the water outside of his body (by merely keeping his blow hole and mouth open), he will feel no physical discomfort from external water pressure as a dive continues. This is due to the fact that the water in his circulatory system and bodily cells, like all liquids, is relatively incompressible. So, we can expect that a whale's deep dive will result in no serious internal injuries occurring to his ribs or other internal organs.

A whale, being a mammal and not a fish, has lungs instead of gills, so he will not be able to extract oxygen from the surrounding water. He will, therefore, be limited in how long he can remain at great depths by the amount of surface oxygen he accumulated in his blood and tissues during the preparatory hyperventilation phase for his dive. Quite possibly, an average sized whale might, by limiting his exertion, be able to remain submerged at maximum depth for up to an hour before he will need to return to the ocean's surface to reoxygenate his blood and bodily tissues. Also, by periodically forcing the sea water out of his lungs and breathing in fresh water, he would, at least, be able to remove some of the carbon dioxide that is produced in his circulatory system by the muscular exertion of his dive.

A human diver must undergo a lengthy decompression phase while ascending from a deep dive because, in order to prevent being crushed by the surrounding water pressure, he has been continually breathing *compressed* air containing about 80% atmospheric nitrogen gas which results in his blood becoming super saturated with this gas. He must pause at various depths during his ascent back to the ocean's surface so that the reduced external pressures there will allow the over supply of nitrogen gas in his blood to be slowly removed via exhalation (note that excess carbon dioxide and oxygen gases are also removed during decompression, but they tend to leave the body much faster than do the nitrogen gas molecules). Once the excess nitrogen gas molecules are so removed, they will not then be able to form bubbles in his circulatory system as the external water pressure acting on this body continues to decrease during his ascent.

In a worst case scenario in which he has not spent enough time at various depths during his ascent to allow the excess dissolved nitrogen gas molecules to leave his body, the bubbles they will form can cause severe cramps and even heart failure as the diver experiences "decompression sickness" which is also known as "the bends" because of the body contorting muscles spasms seen in the condition. Should this condition develop, the diver must immediately be placed into a special decompression chamber at the ocean's surface in which the air pressure can be increased until the pressure acting on his body actually physically collapses the nitrogen bubbles and forces their gas molecules to dissolve back into his blood and tissues again. Once that is done, his symptoms will be relieved and he can then be slowly decompressed at the correct rate in order to finally remove all of the excess dissolved nitrogen gas from his blood and tissues.

However, a whale returning to the ocean's surface after a deep dive requires no lengthy pauses for decompression at various intermediate depths. This is because the whale makes his dive using only the air in his blood and tissues that was originally at sea level pressure and contained a fixed amount of nitrogen gas. Since he does not breathe compressed air continuously while submerged, his blood does not become super saturated with an over supply of nitrogen gas. This then helps suppress the formation of any nitrogen gas bubbles in his circulatory system as he ascends from a great depth in the ocean.

The big question that emerges from this evidence of whale activity at such depths is "what would these generally docile mammals be doing down there?" The answer is rather disturbing. The whales are actually *feeding* on something at those great depths. What they are feeding on are large squid, very large squid!

From an energy conservation point of view, it would just be too laborious for a large whale to be chasing a few small squid around at such cold and dark depths. Rather, it seems far more likely that the whales, using their songs as a form of biological sonar known as "echolocation", are pursuing squid that may have overall body lengths up to and even in excess of two hundred feet.

Such a squid would have a "head" about thirty feet long and ten or so feet wide. Its two beak-like teeth could easily disable or destroy a small manmade submersible vehicle and pose a challenge for hungry whales that hunt them. Squid, like the octopi that they are related to, have eight tentacles each covered with two rows of suction discs. There are also two extra tentacles that are specially evolved to function as whips. These extra whips, known as "feeding tentacles", each terminate in a pad that is covered with powerful suction discs. Each suction disc is further lined with sharp rings of a hard material called "chitin" that actually allow the disc to function like a claw.

These giant squid, technically known as *"Architeuthis dux"* which means "ruling squid", are mentioned in several Old Norse legends (they called them *"kraken"*) wherein they attack Viking long boats and have to be fought off by the crew with swords and axes. In modern times no one has ever seen one alive and the largest dead specimen to wash up on a beach was only about 57 feet long. However, there are several 20th century sightings by Japanese fisherman that suggested a size of about 200 feet, but these reports were only anecdotal in nature.

When preparing to attack its prey, a squid will lash out its whip-like tentacles toward the victim and secure a firm grip on its flesh with them. The prey is then hauled directly toward the squid's razor sharp beak-like teeth which proceed to simultaneously poison and shred the prey as it is consumed. The attack of a hungry squid is so rapid, that most of its victims will be either dead or severely injured before they even know what is happening to them.

Unlike the squid, however, whales, because of their huge brains, are far more intelligent animals and they often use a team approach to accomplishing complex tasks. I can imagine that they would attack a giant squid as a group or "pod" of three. Perhaps the smallest of the three whales swims along slowly as though it is sick or injured and serves as bait to lure a hungry squid out of its hiding place on the ocean floor and into the open.

Most of the animal life in the dark deeper parts of Earth's oceans has the ability to bioluminesce. That is, they can metabolically produce a kind of "cold light" from special cells that they use to identify each other and attract prey to them. Many species of squid also have this ability. However, since the smaller whale being used to draw a giant squid out of its cover does not bioluminesce, we must postulate that the squid can actually somehow visually see it. Perhaps during the daylight hours enough light reaches the gloomy depths of the ocean floor to permit a squid with dark adapted eyes to just barely make out the dark silhouette of the lure whale as he moves along above the squid's position. Of course, if the giant squid bioluminesces, the whales will have no problem seeing him. If the giant squid does not bioluminesce, then the whales will use echolocation to track him.

As the squid draws nearer to the smaller, apparently defenseless whale in preparation for the use of its whip-like grappling tentacles, it will focus all of its attention on its target and not notice the two other whales that are rapidly approaching it from opposite sides (at speeds up to 30 miles per hour). The two attacking whales may then tear out the squid's large, basketball sized eyeballs and thereby immediately render it blind and helpless. Without vision or sonar, the squid can only flounder about aimlessly and all three whales can then proceed to tear it into bloody shreds and consume it. Most likely, however, they do not consume the tentacles or beak-like teeth because these contain sharp surfaces that could injure a whale's mouth or digestive tract.

With their stomachs filled with fresh squid meat, the three whales can then begin a rapid ascent to the ocean's surface while using the remaining oxygen carried in their bloodstreams and bodily tissues. Again we note that, since the nitrogen gas that was originally inhaled along with this limited amount of oxygen at the ocean's surface was not continuously supplied to them at high pressure during their descent, there is no need for the whales to delay their ascent so as to allow for a slow decompression that would be necessary in order to rid their bloodstreams of excess nitrogen gas before it could form bubbles and cause them to experience decompression sickness.

Although it is only a guess, if these huge squid breed at anywhere near the rate observed for their much smaller cousins dwelling at far shallower depths, then there could be several hundred thousands or even millions of them lurking around down there at present!

While human researchers are just starting to explore a very small percentage of the deepest parts of Earth's oceans, there appears to be a substantial body of evidence that this realm has been of particular interest to the various extraterrestrial races that have visited our planet throughout its history. Many highly reliable UFO sightings have occurred at sea and been reported by human witnesses aboard ships.

Sometimes these reports describe mysterious submerged lights that will follow a ship for miles only to suddenly disappear. Such cases are not that frequently mentioned nowadays, but used to be referred to as "USOs" or "Unidentified Submerged Objects" and they are a significant subcategory of the UFO literature.

Occasionally, the water near a ship will begin to violently swirl and dance after which the gleaming metallic hull of a spherical or domed disc UFO emerges just prior to the craft rising up into the sky and then streaking away at incredible velocity. Even more bizarre are the cases in which a UFO actually dives directly in the sea without sustaining any physical damage and then rapidly submerges from view.

Many skeptics of the UFO field use such observations to deny the possibility that a real, physical object was involved in such plunges into a body of water. However, I think that they proceed from a false assumption. Namely, they assume that the UFO, if a physical object, must have had its *normal* mass and weight when it performed such a maneuver.

Indeed, if an occupant carrying vehicle with its normal mass and weight attempted to plunge directly into the ocean, it would be immediately destroyed upon impact with the water's surface. This occurs because the leading surfaces of a vehicle that first contact the surface of a body of water experience a tremendous amount of hydrodynamic drag which then rapidly decelerates the front portion of the vehicle. This drag is equivalent to a suddenly escalating force pushing in the opposite direction that the craft is traveling in and results in the craft's *own* kinetic energy being mostly used to crush its remaining *airborne* structures against its slower moving *submerged* front portion. There are no known cases of earthly aircraft doing such a "nose dive" into the ocean and surviving the impact. They are all totally destroyed upon impact and reduced to smaller pieces of debris. Most of these fragments will sink and a few of the least dense ones will float on the ocean's surface to eventually be carried away by any currents there. Needless to say, the crew of such a craft will also be instantly killed upon impact.

If, however, the craft is totally massless due to the use of anti-mass field generators as I am convinced all *genuine* extraterrestrial UFOs are when airborne, then the situation that results upon an impact with a massive material like the ocean's surface water is very much different (the reader unfamiliar with the concept of anti-mass field generators and their use by UFOs is advised to read the chapter titled "A UFO Propulsion Primer" which appears in my book *The New Science of the UFO*). In such a case, the UFO will, since it is massless, weightless, and without momentum, instantly decelerate to a dead stop upon its front portion making contact with the water's surface. If the craft carries a crew, then they too will also instantly decelerate to a stop. However, lacking any momentum or kinetic energy, neither the craft *nor its crew* will be damaged by such an abrupt deceleration!

Of course, a massless UFO would only come to a complete standstill upon contacting the ocean's massive surface water and the craft will still need some means of forcing its hull beneath that surface and submerging itself. There are two options available to do this.

Upon reaching the water's surface, the pilot of a massless UFO can reduce the rate of emission of anti-mass field radiation from his craft's anti-mass field generators. This action will cause the craft and its crew to regain a portion of their normal mass and weight. This will be sufficient to then cause the UFO to submerge if the regained mass of its various structures exceeds the mass of the volume of water displaced by those structures which would include a watertight crew section. After such a craft was underwater, it would then have to rely upon some sort of mechanical (i.e., propellers) or, if in salt water containing various ions, electromagnetic propulsion system to provide a thrust needed to push its then partially massive hull and crew along. Craft that do this will not be able to travel any faster underwater than do earthly submarines.

Alternately, some UFOs, which have become USOs, may continue to remain completely massless and weightless while underwater. It is these craft which are capable of performing bizarre maneuvers wherein they appear to literally plunge directly into a body of water and then quickly disappear from view as they submerge.

In these cases, the massless craft will invariably be using the plasmadynamic mode of propulsion to generate thrust when airborne. Basically, the craft is able to project crossed magnetic and electric fields whose field lines are perpendicularly oriented with respect to each other beyond its hull's surface from special devices I refer to as "drive units" that are concealed beneath its *non*-ferrous metallic hull. As the craft's "excess" anti-mass field radiation that is not needed to completely negate the mass of the craft and its crew leaves the outer surfaces of the vehicle's hull, it interacts with the projected magnetic fields immediately surrounding those surfaces and is converted into a form of anti-mass field radiation that has the ability to *reduce* the normally present *attractive* electrostatic forces that act between electrons and their nuclei and which hold them inside of their individual atmospheric gas atoms.

Once this happens, the ambient thermal energy of the air itself becomes sufficient to knock the loosened outermost electrons off of their respective atoms. The result is that the layer of atmosphere immediately surrounding the massless UFO's hull will become highly ionized and converted into a rich boundary layer of plasma. The craft using the plasmadynamic mode of propulsion for thrust then uses its drive units' crossed electric and magnetic fields that project out into this surrounding layer of atmospheric plasma to apply what are known as "Lorentz forces" to its various ionized gas particles.

These forces make the plasma particles flow at very high velocity around the craft's hull *without* touching it so that a condition known in the science of aerodynamics as "laminar flow" is achieved. This is how a massless UFO can move so rapidly through the dense lower portions of a planet's atmosphere without producing a "sonic boom" even though the vehicle's velocity can greatly exceed the speed of sound. As the virtually massless plasma particles are forced to flow around the moving UFO's hull at high velocity, some of them can emit electromagnetic radiation in the visible portion of the spectrum which is known as "cyclotron radiation". This visible radiation accounts for the bright amorphous glows often seen to envelope nocturnal UFOs and, when unusually bright, even daylight UFOs.

With this basic understanding of how plasmadynamic propulsion systems operate, it is a simple matter to explain how a massless UFO using such a system for thrust could be able to survive a direct plunge into the ocean's surface a high velocity without this action causing any damage to the craft or its crew.

In this maneuver, the UFO simple makes the transition from forcing highly ionized atmospheric gas particles to flow around its hull to forcing highly ionized *water* molecules to flow around its hull. Upon contacting the water's surface, water near the craft's leading surfaces will be rendered nearly massless there by the emission of the craft's mass reducing anti-mass field radiation. The atoms of this surface water's molecules, like those atoms of the atmospheric molecules still near the unsubmerged portions of the craft's hull, will then also become highly ionized by the process described above and be forced by the Lorentz forces produced by the vehicle's drive units to begin flowing around the submerging leading surfaces of the UFO's hull without touching it.

Thus, the massless craft will feel no hydrodynamic drag forces acting on its hull as it enters the water's surface. However, the craft will slow upon contact with the water because its plasmadynamic drive units can not move the much denser water around the hull as quickly as atmospheric plasma can be moved. But this reduction in velocity will probably not be that apparent to outside observers. To them, the craft will appear to have made a smooth and instant plunge into the water's surface. Because the craft and its contents are massless at all times, they actually have no kinetic energy and their reduction in velocity upon suddenly slowing and submerging will not give rise to any inertial forces that could cause damage to the vehicle's hull or its crew.

Once underwater, however, a massless USO will have a problem that a craft with mass will not have. A craft with most or all of its full normal mass can rely upon its weight to counteract the lifting force of buoyancy on its external hull and thereby keep itself submerged. A massless and weightless vehicle, on the other hand, will experience a strong buoyant force acting upon its hull which will try to force it to surface. Like all submerged objects, a massless USO's hull will be buoyed up by a force equal to the weight of the volume of water displaced by its hull. In other words, despite the fact that a submerged USO might be completely massless, the craft is still subject to Archimedes famous principle of buoyancy.

It is also important here to realize that any USO being used to explore the deepest portions of a planet's oceans must, just like an earthly deep sea submersible vehicle, have a hull that is capable of withstanding the enormous external water pressures encountered at those depths. If such a craft exceeds the maximum depth whose pressure its hull can withstand, the craft's hull will be damaged and may even rupture, flood out, and result in the occupants being killed. Massless vehicles are not immune to the effects of either external atmospheric or water pressure.

In order to remain submerged, a massless USO will have to continuously use its drive units to force ionized water to flow from its lower hull surfaces to its upper hull surfaces. This action will cause water molecules to pile up above the craft's upper hull surfaces and the pressure from this congestion will result in a force that pushes down on the hull. This downward acting force can then be carefully adjusted so that it perfectly counter balances the rising force created by the USO's natural buoyancy and the craft can then remain continuously submerged.

When the craft's pilot wants to descend to greater depths, he simply increases the rate at which ionized water below his craft is forced to flow around its hull and toward its upper surfaces. The increased water pressure acting on the upper surfaces of the massless USO then causes the downward force acting on the craft's hull to be greater than the buoyant force which is always acting to cause the hull to rise and the vehicle will descend. Ascending is achieved by merely reducing the rate of flow of ionized water moving from the craft's lower to upper hull surfaces so as to reduce the quantity of water molecules that pile up above the upper hull surfaces. At some point the downward force acting on the USO's upper hull surfaces will be less than the buoyant force acting on the craft's lower hull surfaces and the USO will begin to ascend.

As a massless USO reaches the surface of the water, surface witnesses, should they be close enough, may momentarily see a rather strange sight.

If the craft is disc shaped with a domed upper surface, nearby witnesses may see a large "bulge" of water appear and then momentarily rise up above the surface of the water as the USO begins to emerge from beneath the body of water's surface. This bulge is actually the boundary layer of virtually massless water that envelopes the craft's curving upper hull surfaces and is carried along with the emerging hull.

As the disc-shaped craft continues to emerge, its outer rim will eventually reach the surface of the water. At this time the drive units located just beneath the metallic surface of the rim can momentarily force outside water near the rim to spray upward to create a fountain-like effect at various locations around the rim. This is because the projected and crossed electric and magnetic fields emanating out of the rim's drive units and into the surrounding water are still trying to force ionized water from beneath the USO to flow to its upper hull surfaces in order to control the craft's rate of ascent even though those surfaces are now exposed to the air. Because the water near the craft's outer rim is still almost completely massless due to its proximity to the massless hull, it can spray upward several feet before it is far enough away from the mass negating influence of the craft's emitted excess anti-mass field radiation so that it can then regain enough of its original mass and weight and begin falling back down toward the water's surface.

Once the craft is out of the water and airborne again, the pilot will then begin engaging its propulsion system's other drive units which are located below its hull in various locations so that they can force ionized air or atmospheric plasma to begin flowing around the hull for thrust in a direction that will rapidly take the vehicle away from the water's surface. As these drive units are used, the bulging layer of water that still covers the craft's upper domed surface and which can be several inches thick will then break open to form an expanding circular hole at its center as it is pushed toward the rim by the incoming rush of Lorentz force driven atmospheric plasma. As the craft's gleaming metallic hull finally becomes fully visible in the daylight, any water that previously surrounded its emerging hull will just fall back into the body of water.

As the massless craft, once again called a UFO since it is now airborne, continues to rise, all of the water that surrounded its previously submerged metallic hull will finally fall back into the ocean. The vehicle's pilot can then reverse the directions that his craft's rim drive units will force nearby and nearly massless ionized atmospheric atoms and molecules to flow around the craft's hull (this is done by simply changing the directions of the electric fields these drive units project out beyond the hull and into the surrounding boundary layer of atmospheric plasma particles). This action will then apply Lorentz forces to the atmospheric plasma ions formed near the craft's upper hull surfaces and make them rapidly flow downward at various locations around the rim to the now airborne UFO's lower hull surfaces. Note that not all of the atmospheric molecules will be ionized by a UFO's plasmadynamic propulsion system and most will still be electrically neutral. However, the motion induced in the nearly massless ionized air molecules will tend to also force the also nearly massless uncharged molecules it contains to also begin flowing rapidly around the hull. This then results in a build up of atmospheric pressure on the craft's lower hull surfaces that rapidly propels the massless vehicle skyward.

As the massless craft rises into the sky, other drive units hidden within its nonmagnetic hull surfaces can then be brought into operation by the pilot in order to allow the UFO to engage in controllable horizontal flight. These drive units will then apply horizontal driving forces to the plasma particles surrounding the hull so that their direction of flow will cause a pressure differential to develop between the craft's leading and trailing surfaces. The force created by such a sustained differential will then drive the massless vehicle forward toward whatever direction that its leading surfaces face. Of course, while this is going on the pilot will be viewing the external scenery through small cameras built into the hull of his craft and can use that visual information to make whatever course adjustments are necessary in order to fly the UFO toward a selected destination. Most UFOs seem to employ this indirect method of viewing for the pilot to use rather than equipping the craft's hull with windows through which he could directly view his surroundings. The reason is probably because using transparent windows increases the risk that a shattered pane would result in the crew section undergoing "explosive decompression" that could quickly kill a crew not wearing spacesuits. A completely closed hull avoids this risk.

Although it is more difficult to keep a massless USO underwater than a massive one, there are some distinct advantages that massless USOs have over massive USOs. Mainly, the massless USO can travel much faster underwater than a massive USO. Since sea water is about 1000 times denser than air, we can roughly expect a plasmadynamically driven massless UFO to be able to achieve about $1/1000^{th}$ of its maximum airborne velocity when it operates underwater and becomes a USO. (The reason for this being that, for given intensities of their projected electric and magnetic fields, the vehicle's drive units can only move a somewhat fixed *quantity* of ionized particles and any electrically neutral particles swept along with them through the boundary layer surrounding the hull in a given amount of time. If the density of the medium surrounding its hull is suddenly *increased* by a certain factor as happens when a craft leaves a planet's atmosphere and submerges into its ocean, then, in order to keep the quantity of particles moving through the boundary layer constant, the velocity of these particles must be *decreased* by the same factor.) So, if such a craft could achieve a maximum velocity of 10,000 miles per hour while airborne, then we can expect it to able to move at about 100 miles per hour while underwater.

This is a very rapid when compared to the submerged velocities of such objects as earthly submarines. Also, while underwater, a massless USO's drive units may cause the ionized water molecules it forces to flow around the hull to emit visible cyclotron radiation so that these craft can glow while submerged just as they can while airborne. Other than sonar detection, such visible glows are usually the only indication that a surface human observer has of the presence of such a submerged extraterrestrial vehicle.

As mentioned above, a USO, whether massless or massive, will still have to contend with the problem of increasing water pressure as it descends to greater depths in a planet's oceans. This will require an external hull which is not only water tight, but also capable of withstanding the increase in external water pressure during a deep dive. Since most UFOs use metallic hulls that may be an inch or more in thickness, we can expect them to be able to readily descend to the depths most earthly military submarines can achieve. This would be in the range of about 1000 to 2000 feet. Beyond that depth, a disc-shaped craft becomes difficult to reinforce against external water pressure and the use of cylindrical and spherical vehicles provide a structurally stronger and more reliable design. To reach depths in the 10,000 to 20,000 foot range will require the use of a spherical hull and it will have to have a thickness of a foot or more just like the steel spheres that enclose human oceanic explorers using earthly bathyspheres and bathyscaphs.

By now it is fairly safe to assume that most of the regular extraterrestrial visitors to our planet from local star systems (being, perhaps, within a distance of 50 light years from ours) have conducted a complete survey of our oceans and the life they contain. Exploring the deepest oceanic trenches that drop to depths of four miles or more, they have, no doubt, noted the huge variety of life down there. Occasionally, one of their USOs may be subject to attack by some gigantic form of sea life such as the huge squid described previously. Like the fictional Nautilus submarine of Captain Nemo, it would seem that the best way to defend their craft would be to use an electrical defense system.

As a giant squid attempted to consume a spherical USO, the submerged craft's pilot would be able to electrify various sections of his vehicle's hull and, thereby, deliver a voltage of tens of thousands of volts into the tentacles of the attacker. This voltage would not, of course, be intended to kill the squid, but, rather, merely to force it to immediately release the USO and jet propel itself away in terror. Indeed, such a defense system would seem to be automatically provided by any massless USO that employs its propulsion system's drive units in order to stay submerged and move about horizontally. This is because the electric fields projected into the surrounding highly ionized boundary layer of sea water emanate from regions of different electrical charge induced on the craft's metallic hull by oppositely electrically charged capacitor plates located just below the hull's surface that are insulated from the hull. The voltage difference between neighboring charged regions can be in the thousands or even tens of thousands of volts and would deliver a painful electrical shock to any squid attempting to wrap its tentacles around the hull as he tried to consume the craft.

The physical capturing of a 200 foot long squid would probably not be of interest to visiting extraterrestrial beings although such a task would certainly be technically possible for them. They would probably be content to merely record images of the creature from a distance. However, it is entirely conceivable they might capture smaller bioluminescent sea organisms to study and return to their home worlds for display. It might even rarely be the case that they transplant various species of fauna and flora from other planets into our oceans to see if they can adapt to a new environment. Such occasional transplants might account for the recent discovery of certain plants that have been found in active underwater volcanoes and which are actually able to feed off of the various toxic gases escaping there. Since there are no other plants on Earth with this ability, some have speculated that these very unique plants may have been brought to our planet aboard meteorites or comets. I am of the opinion, however, that it is far more likely that these transplanted species were either accidentally or deliberately released into our oceans by extraterrestrial visitors.

Alien exploration of Earth's oceans is also probably not limited to their examination of aquatic organisms and geological features. The cover of water can provide extraterrestrials with a perfect opportunity to examine some of the technological products of humanity that lie rusting out on the sea floor. Sunken ships, submarines, and the wreckage of ditched airplanes are there for their close inspection. Ufonauts (perhaps they should more properly be referred to as "usonauts") could directly enter these wrecks and then gain an in-depth knowledge of Earth's technology from various periods of human history. Or, they could use remote unmanned probes (actually just miniature USOs) to more conveniently and safely explore and analyze the scattered debris of our fragile technology.

Finally, let us consider what interactions can occur when our extraterrestrial visitors encounter single ships in the middle of our oceans.

In most cases, the UFO's crew will merely examine the vessel and then be on their way. Occasionally, however, a UFO will hover close enough to a ship so that the UFO's propulsion system causes the ship's lighting system to dim and engine to fail. These mysterious effects are known as "EM effects" in ufology and are apparently due to an enhancement of the Edison Effect inside of the evacuated glass envelopes of the ship's various incandescent lamps and the creation of electrically conductive atmospheric plasma near the high voltage ignition circuitry of the of ship's internal combustion engine (I dealt with these effects extensively in my first UFO book, *Secrets of UFO Technology*).

As happens during the production of the rich atmospheric plasma immediately surrounding the hull of a plasmadynamically driven massless airborne UFO, the EM effects earthly seagoing ships experience are the result of some of the excess anti-mass field radiation emanating from a nearby massless hovering UFO interacting, after traveling a short distance, with the magnetic fields found inside of a ship's incandescent lamps and engine compartments. As in the case of such similar failures induced in automobile lightbulbs and ignition systems, the effects are temporary and quickly diminish after the UFO departs the scene. However, it is interesting to note that the oceanic electrical failures generally do not take place when there is a *submerged* USO near the earthly ship's location. In these cases, the denser water surrounding the massless USO prevents any of its emitted excess anti-mass field radiation from reaching the surface ship's electrical system components and causing them to fail.

On rare occasion, however, an airborne UFO at sea may have a more intimate interaction with the crews of small earthly sailing vessels. If the interaction occurs at night, those crew members on the ship's night watch may be quietly abducted and given medical examinations while other crew members are unaware or asleep. Because of the desolate situation of an earthly ship at sea, this is a unique opportunity for extraterrestrials to examine human beings without fear of discovery and interference.

Finally, we must consider another bizarre phenomenon for which there seems to be some support in the literature devoted to the mysteries of Earth's oceans. I am here referring to the many cases on record of ships that have been found drifting aimlessly about on the open seas and which have been abandoned by their crews. Another ship then arrives on the scene and sends a boarding party over to inspect the deserted vessel. In many cases the abandoned ship's mess or dining area will actually have meals left out that are still warm! Usually the vessel's cargo is intact while one or more of its lifeboats are missing. Yet, a scan of the ocean's horizon in all directions shows nothing. The impression the boarding parties get in these cases is that the abandoned ship's crew left in hurry, but for reasons that are unknown.

It would seem that in these cases of mysterious crew disappearances at sea the crew must have been airlifted off of the abandoned ship. Since many of these cases occurred in the 18th and 19th centuries, the possibility of earthly aircraft being responsible is eliminated. One is therefore forced to the conclusion that the missing crews of these ships were abducted by extraterrestrials that had no intention of returning them!

Such abductions of whole crews consisting of a dozen or more humans can only have one implication. Just as our extraterrestrial visitors transplant lower life forms between planets to study how they adapt to new environments, so, too, they must occasionally transplant higher lifeforms between different worlds to observe how they also adapt to a new environment. The humans taken from our ships at sea were probably taken to Earth-like planets in other star systems of our own galaxy in order to determine how well our species can adapt to a new environment. Quite possibly, to eliminate the psychological trauma of such a transplant, the long term memories of the humans abducted would be erased by the use of various drugs. This process would, most unfortunately, probably also erase any technical expertise that each human had acquired during his lifetime on Earth, but his basic ability to acquire new skills would still be intact.

On their new world, the human abductees would form a simple society and pool their abilities to ensure their mutual survival. The introduction of female humans would allow for population growth and continuing technical progress. Perhaps humanoids from other *non* spacefaring cultures similar to that of Earth would also be deposited on the host world to see how they developed and interacted with the humans there. In time, an entire planetary culture composed of various transplanted humanoids could be constructed. As their biological sciences developed, they would eventually realize that their civilization was not a natural one, but rather one that had been artificially created.

Such experiments using sentient beings surely seem to be cruel ones, but the extraterrestrials that perform these types of abductions probably make an effort to select humans who would have the least social attachments on their home worlds and whose sudden and mysterious disappearances would be dismissed as a tragic, but not completely unexpected occurrences considering the hazardous environments in which they live and work. Thus, the aliens would seek out solitary individuals located in Earth's various desolate areas such as the middle of our oceans. The people abducted would generally be excellent physical specimens in terms of their health and would be intelligent enough to acquire and use the skills needed to survive in a hazardous environment. In short, they would be ideal for an interstellar environmental transplant experiment.

I can not, of course, condone such experimentation. I can only hope, however, that any knowledge gained from it is eventually put to some moral use such as helping the most primitive of humanoid beings in danger of extinction on other planets in our galaxy survive, thrive, and then go on to develop their own independent planetary cultures.

Chapter 3

The Hidden Realms of the Underworld

ONE CAN SPEND SO MUCH time contemplating the possibility of extraterrestrial life which accounts for the various UFO sightings reported throughout history that one may overlook another equally exciting possibility. I am referring to the possibility that our planet is also host to other forms of sentient life apart from its surface dwelling human populations. Indeed, after a lifetime of studying bizarre phenomena and sightings, I am somewhat convinced that the upper layers of Earth's rocky crust are inhabited by subterrestrial beings who are actually the very remote ancestors of its surface humanity!

I am of the belief that many hundreds of thousands of years ago as the various Ice Ages caused the Earth's polar caps to periodically expand toward and then retreat from its equator, early surface dwelling humans became split into two distinct groups: those who could use fire and those which did not possess this ability. The fire makers would have been able to survive the freezing surface temperatures by using their incendiary skills to heat their drafty caves or huts, but the non-fire makers would have been forced to seek out the natural geothermal heat within the Earth's crust in order to survive. These non-fire makers would have escaped the frigid ice flows by moving into natural limestone caverns and then, via the crevices found there, descending farther down into the crust where temperatures are higher.

Most people do not realize that the ambient rock wall temperature inside of the Earth's crust rises by about 1.8° Fahrenheit for every 100 feet that one descends into the planet (and any air at that those depths also acquires the same temperature if it is not exchanged with air from other depths). If a surface cave's ambient rock wall and air temperatures were 70° F, one would have to descend to a depth of about 1666 feet before the temperatures in a passageway reached 100° F. If the Earth's surface temperatures were reduced to the freezing point of water or 32° F by a covering layer of glacial ice, then it becomes possible to descend to a depth of 3777 feet or about 7/10ths of a mile before the ambient rock wall and air temperatures reach 100° F. And, of course, even colder surface temperatures would result in greater depths being attainable before the temperature any subterranean dwelling organism was exposed to reached 100° F.

Thus, we see from this that modern humanity's early non-fire making ancestors could, at a minimum, have easily used natural caverns and crevices to explore the Earth's crust to considerable depths since 3777 feet down into the Earth is a distance equal to about 3 times the height of the Empire State Building!

However, these early subterrestrial dwelling humans would have been exposed to something at those depths for which they would have been unprepared by their slow surface evolution: increasing levels of various forms of radioactivity that occur due to the natural radioisotopes contained in the very rocks and minerals of the natural and artificial passageways that they would have used. Ordinarily, a surface dwelling human being experiences a whole body exposure to about 125 millirems of radiation per year from all natural sources. Our subterrestrial dwelling ancestors, however, may have had exposures to *thousands* of millirems per year. Such continuous exposure can produce a chronic form of low level radiation poisoning and result in genetic damage that shows up in the form of an increased rate of birth defects and miscarriages along with various cancers.

The early subterrestrial humans would, however, have had to endure these radiation related problems because they would be preferable to the certainty of death by freezing on the Earth's frigid surface. Although live births would be the exception, we can imagine that, over the course of tens of thousands of years, these beings would have evolved the ability to withstand their highly radioactive environment. With an enhanced rate of genetic mutation and natural selection taking place, we can imagine that those living deepest inside the Earth's crust might have developed thick skins which would contain high concentrations of various metals that would effectively stop or, at least, greatly reduce the amount of radioactive decay particles and ionizing radiation that penetrated deeper into their bodies. Depending upon the particular organometallic compounds in their skin, these beings might have a greenish or bluish skin color when viewed in white light. Thus, the initially elevated rate of mutation would, as long as it did not kill an entire subterrestrial population, actually have allowed them to relatively quickly evolve the ability to adapt to their new environment.

Their rapid rate of mutation would also have allowed them to evolve other capabilities that would be essential for survival at depths of up to a mile or so underground. For example, the limited supply of air in underground chambers would prohibit the use of fire for illumination and, as a result, all of the subterrestrial beings would eventually develop the ability to see in the infrared region of the electromagnetic spectrum. Warm rock wall surfaces and any surface objects they brought along with them such as weapons or tools would be readily visible to any being able to see his surroundings via infrared radiation.

Initially, their adult males would all have been muscular individuals evolved to successfully hunt prey on the Earth's surface while their females would have had the strength needed to nurture offspring and defend them against the occasional predator. But, once confined to their subterrestrial realms, both genders would, over the course of successive generations of radiation induced evolution, have become smaller as this new body size was preferred so that they could then more easily squeeze their way through various natural passageways that tended to grow narrower with increasing depths.

To facilitate rapid movement through tight passageways and crevices, it is desirable to have a low body mass and weight. To obtain this characteristic, rather than continue to shrink in size, they would eventually have evolved the ability to *voluntarily* alter their normal body masses and the gravitational and inertial properties associated with them. Thus, they would retain their normal bodily masses and weights which were needed to walk easily about the level floor of a large subterranean cavern in a standing position or could greatly reduce them almost to extinction when the need arose to be able to safely float from one level to another or pull themselves along, hand over hand, through a long and cramped channel in a thick rock wall that separated two adjacent caverns. Eventually, if needed for survival, they would have evolved the ability to become completely massless at will.

From studying the few well documented cases of human beings who were able to levitate themselves, I am convinced that the natural physiological process by which this is achieved is very similar to the artificial process that is utilized by an extraterrestrial craft or UFO in order to negate its and its crew's normal mass and weight so that they can hover in Earth's atmosphere without any apparent means of propulsion and, when in space, achieve faster than light travel velocities. This process, when used by a levitating human, requires the emission from his body of a form of *non*-electromagnetic radiation which I call "anti-mass field radiation". (In my later research I realized that this radiation is actually corpuscular in nature and consists of a continuous outward flow of huge quantities of long "streams" of *antigravitons* in which the antigravitons within any single stream are in contact with each other as are the spherical beads on a necklace. The interested reader can find a far more detailed treatment of their properties in my three previous volumes devoted to the UFO phenomenon.)

This radiation, in the cases of the *very* few humans who could naturally emit it in quantities sufficient to result in their bodies becoming light enough to achieve atmospheric buoyancy, was produced by the interaction of electric and magnetic fields that could temporarily form within their circulatory systems. Once emitted, anti-mass field radiation basically neutralizes and thereby cancels out the normally present "mass field radiation" that emanates out from all of the mass possessing subatomic particles that compose the atoms of such a rare person's body and which give rise to his normal properties of weight and inertia.

My later research eventually convinced me that mass field radiation is corpuscular in nature and consists of a continuous outward flow of huge quantities of long streams of *gravitons* that, again, within each stream are in direct contact with each other as are the beads on a necklace string. Antigravitons and gravitons behave somewhat like antiparticles of each other such that when a single antigraviton stream comes close enough to and then aligns lengthwise with a single graviton stream, the two streams rapidly fall apart as each of their individual particles are attracted to and then form a tight two particle "couplet" with one antiparticle in the *adjacent* stream. These couplets that continuously form then just fly off into space at light velocity with the result that the mass of the object emitting the mass field radiation will be lowered as its normal gravitational and inertial properties are lost.

Each of these kinds of field radiations can be either artificially or naturally produced although it is usually only the anti-mass field radiation that is either artificially or naturally produced.

In a massless airborne UFO the anti-mass field radiation is almost always produced artificially by the craft's propulsion system although a very small percentage of UFOs may actually be space traveling biological organisms that emit anti-mass field radiation from their circulatory systems. In the body of a levitating human anti-mass field radiation is always naturally produced by his circulatory system. For the occasional case of a UFH or "Unidentified Flying Humanoid" that is seen levitating at a distance from his extraterrestrial craft, the anti-mass field radiation that permits this is almost always artificially produced by equipment that he has strapped to his body although some UFH may have evolved the ability to levitate using the emission of anti-mass field radiation from their circulatory systems.

From my research into UFO technology, I learned that a UFO's emitted anti-mass field radiation can be used for more than just negating it and its crew's normal mass and inertia. For example, by unbalancing the intensities of the electric and magnetic fields used to generate the craft's anti-mass field radiation, it is possible to emit anti-mass field radiation with different types of "character". There are *three* distinct types of character possible for this radiation and two of them have opposite effects on the electrical properties of any matter they penetrate.

The first character type for anti-mass field radiation I refer to as "neutral". This is the type of anti-mass field radiation that is usually used by massless airborne UFOs and levitating humans and which is responsible for negating their normal mass, weight, and inertia. Neutral anti-mass field radiation is produced whenever the intensities of the electric and magnetic fields that are used to produce it are balanced in intensity with respect to each other and this is why I selected the term "neutral" to emphasize that neither the electric field nor the magnetic field predominates when this character type of anti-mass field radiation is produced.

Neutral anti-mass field radiation has no effect upon the electrostatic forces that act between the electrically charged subatomic particles found in atoms. These electrostatic forces act between an atom's negatively charged electrons and also between its negatively charged electrons and its positively charged nucleus (which is positively electrically charged because it contains one or more positively electrically charged protons). These forces hold atoms together and are responsible for giving the atoms of each of the 115 different chemical elements a distinctive atomic structure which, ultimately, determines all of the unique chemical, magnetic, optical, and physical properties of that element.

The second type of anti-mass field radiation character I refer to as "E rich" because the *electric* fields involved in its production are stronger in intensity than are the magnetic fields. This type of anti-mass field radiation tends to weaken the electrostatic forces of *attraction* acting between *unlike* electrically charged subatomic particles in atoms such as exist between an atom's positively charged nucleus and the surrounding "cloud" of negatively charged electrons that orbit the nucleus at high velocity. Since the symbol "E" is often used in physics to denote the intensity of an electric field, I chose to use it along with the adjective "rich" to emphasize the predominance of the electric field's intensity when anti-mass field radiation with E rich character is produced.

The third type of anti-mass field radiation character I refer to as "B rich" because the *magnetic* fields involved in its production are stronger in intensity than are the electric fields. This type of anti-mass field radiation tends to weaken the electrostatic forces of *repulsion* acting between *like* electrically charged subatomic particles in atoms such as exist between the positively charged nuclei of two neighboring atoms or which exist between their negatively charged electrons. Since the symbol "B" is often used in physics to denote the intensity of a magnetic field, I chose to use it along with the adjective "rich" to emphasize the predominance of the magnetic field's intensity when anti-mass field radiation with B rich character is produced.

In the case of our planet's subterrestrial beings which we are considering in this chapter, the ability to voluntarily alter the character of their bodies' naturally emitted neutral anti-mass field radiations would have evolved slowly over time. The use of these three different characters of anti-mass field radiation then gave them some truly amazing paranormal abilities that aided their continued survival in the underworld realms in which they dwelt. Let us now consider what these abilities are and how they are employed.

As previously mentioned, E rich anti-mass field radiation, which weakens the electrostatic forces acting between *unlike* electrically charged subatomic particles, will reduce the attraction between the positively charged nuclei in the atoms of a subterrestrial being's body and the negatively charged electrons of those atoms. This effect then causes all of the electron orbits of the being's atoms to expand their radii slightly and this then causes the *differences* between the energy levels of *any* two orbits within an atom to decrease by a small amount. It is these tiny energy differences that determine what frequencies of electromagnetic radiation an atom will interact with as incoming photons of electromagnetic radiation are absorbed by the atom and cause some of its electrons to jump from lower orbits to higher orbits that have larger radii. It is the frequencies of the various electromagnetic radiation photons absorbed by his atoms which ultimately determine the colors and other optical properties of a subterrestrial's bodily tissues.

If a subterrestrial creature can emit enough E rich anti-mass field radiation, he can actually force his bodily atoms to no longer interact with electromagnetic radiation photons in the *visible* region of the electromagnetic spectrum. When that happens, he will become invisible as far as these photons are concerned! If he slowly increases the intensity of his emission rate of the E rich anti-mass field radiation until he reaches the rate necessary for complete optical invisibility, he will seem to slowly fade from the view of outside observers who can still only see in the visible region of the spectrum. Should the subterrestrial be able to rapidly build up his emission rate to the necessary level, he can then actually appear to almost instantly disappear from the view outside observers!

The ability to become invisible can be a powerful means of defense as well as offense. It allows one to easily escape from enemies or to be able to sneak up on them undetected for an attack. When the subterrestrials first entered the crust of the Earth many hundreds of thousands of years ago, they must have had to compete with each other for space and food. Those that evolved the ability to become invisible first would have had a tremendous advantage in their ongoing struggle for survival.

There is a problem with being able to achieve optical invisibility, however. Once photons of visible light can pass through the retinal cells of a subterrestrial being's eyes without being absorbed and affecting them, he will need to have retinal cells that can interact with photons of lower frequency. He will need to have retinal cells that can be stimulated to form images by using incoming photons of infrared radiation. Apparently, when such a being's bodily atoms no longer can interact with photons of visible light, they then automatically begin to respond to photons at lower frequencies. Thus, as the subterrestrials evolved the ability to become invisible, they would also have simultaneously evolved the ability to see in the infrared region of the spectrum. A subterranean cavern that would appear pitch black to a surface human limited to only seeing in the visible region of the spectrum would, depending upon its ambient air and rock surface temperatures, appear brightly lit to such a creature. Indeed, the very air in the cavern would have a bit of a glow to it, but not enough to obscure his seeing the rocky surfaces of the cavern.

Most likely, our planet's subterrestrial creatures first evolved the ability to only allow their eyes to become optically invisible so that their retinal cells would respond to incoming photons of infrared radiation. Then, slowly over time with continuing environmental radiation induced genetic mutations, this ability spread to include their entire bodies. Those able to achieve and willing sustain complete bodily invisibility would easily survive the struggle for survival and then pass on the mutated genes for their abilities to their offspring.

As their small populations continued to grow and evolve, these subterrestrial beings would naturally tend to explore crevices and passageways deeper within the Earth's crust in search of additional cavern living spaces. Eventually they would reach depths where the ambient air and rock wall temperatures began to rise well above 100° F. With increasing temperatures, the rock wall surfaces and the air in the caverns at these depths would begin to emit huge amounts of infrared radiation at the higher frequency end of the infrared region of the spectrum of electromagnetic radiation.

This radiation tends to be primarily absorbed by the electrons involved in the chemical bonds that form between the atoms in molecules and causes portions of those molecules to begin violently spinning and vibrating. It is this increased molecular motion that then causes the larger structures made of these molecules, such as bodily cells and the tissues they form, to begin heating up when the molecules are exposed to infrared radiation. When photons of such higher frequency infrared radiation begin to be absorbed at rates beyond a certain level by a subterrestrial being's body, the molecules in the cells of the tissues exposed to them will actually begin to be torn apart and the highly reactive fragments that result, known as "free radicals", can then join together to form molecules not normally found in the cells which either serve no functional or structural purpose in a cell or are actually toxic to it. If the level of radiation exposure is excessive, then the being whose body contains the damaged cells will die.

To survive the rising temperatures at these farthest inhabitable depths by keeping their bodily cells from overheating and being killed, the subterrestrials would, interestingly, be obliged to evolve the ability to make their bodies further invisible to the absorption of even these higher frequencies of infrared radiation photons. They would still be optically invisible and able to see in the infrared region of the electromagnetic spectrum, but their bodily cells, including their eyes' retinal cells, would then only be interacting with lesser quantities of infrared photos at the *lowest* frequencies of the infrared band of the electromagnetic spectrum. These photons carry much less energy and would cause far less heating of their bodily cells when absorbed by the molecular bonds found in those cells.

Once subterrestrials could withstand the hottest temperatures within the deeper portions of our Earth's crust (which is only about 40 miles thick and literally floats on a huge molten sea of material called "magma"), the only major limitation that they would encounter would be due to the decreasing sizes of the crevices and passageways that they would have to travel through as they attempted to move horizontally from one cavern to the next. These narrowing channels would, perhaps, be encountered at depths of about a mile or so below the Earth's surface and would prevent all but the smallest of these creatures from traveling through them.

The problem of rock barriers to mobility at these depths is, amazingly enough, neatly solved by the *other* distinct character that the anti-mass field radiation emitted from a subterrestrial's circulatory system can have depending upon the imbalance in the intensities of the electric and magnetic fields in his circulatory system that produce the anti-mass field radiation. I am referring to the B rich anti-mass field radiation that these subterrestrials can voluntarily produce and which can then greatly reduce the electrostatic forces of *repulsion* that normally exist between *like* electrically charged subatomic particles in matter. By emitting anti-mass field radiation of this character, a subterrestrial being could, at least momentarily, be able to interpenetrate solid rock!

As most technically oriented readers of this chapter will be aware, solid matter is really over 99.99% empty space and the only thing that prevents any two pieces of solid matter from passing directly through each other are the powerful forces of electrostatic repulsion that arise between the atoms of both pieces as their like charged electrons and their like charged nuclear protons approach each other. If, however, as I believe most of Earth's subterrestrials can do, one could greatly reduce these electrostatic repulsive forces that normally arise, say, when a hand makes contact with a solid rock wall, then one could easily push one's hand and even the whole length of his arm directly *into* the surface of the wall of rock and then even deeper into its interior!

Since the structural integrities of the hand, arm, and rock material are all *separately* maintained by the electrostatic forces of *attraction* that exist between the *unlike* electrically charged subatomic particles in their individual atoms and molecules (that is, between their atoms' negatively charged electrons and positively charged nuclei), each continues to possess its normal chemical and physical properties despite the state of interpenetration in which it exists.

A person whose hand and arm had, while voluntarily emitting the B rich anti-mass field radiation from their circulatory system needed to greatly reduce the normally present repulsive electrostatic forces present, interpenetrated a solid rock wall would still be able to flex his fingers while they were in the rock material and then eventually withdraw this hand and arm from the rock without in any way affecting the rock or harming his hand and arm. Of course, if something went wrong with his metabolic processes that produced the B rich anti-mass field radiation being emitted from his hand and arm or somehow blocked the flow of blood through them, then these body parts would no longer interpenetrate the rock but, rather, would actually fuse with and become part of the rock's structure! In such a case, the circulation in his limb and hand would immediately stop and their cells would quickly die from lack of oxygen.

Obviously, using B rich anti-mass field radiation to achieve interpenetrability is somewhat dangerous should a failure in its production occur at the wrong time. Thus, this ability needs to be used as briefly as possible in order to minimize the risks involved with it.

As was noted above, the ability to momentarily interpenetrate solid rock would allow some of the more adventurous subterrestrials to explore and inhabit the deepest caverns in the Earth's crust. They could use this ability to push their way through the tightest of crevices and passageways. Most likely, they would even use this ability to occasionally push their entire bodies through rock walls that might be several feet thick in order to reach the open space of a cavern on the other side.

If the reduction in the repulsive electrostatic forces acting between a creature's body and the rock material is not total but they are just very greatly reduced, then there will be a small amount of resistance force or drag acting on the subterrestrial's body as he pushes his way through a rock wall that is several feet thick. In fact, by carefully adjusting the ease of interpenetrating the rock material by voluntarily adjusting how much B rich anti-mass field radiation his circulatory system emits, he might actually be able to "swim" through tens of feet of solid rock until he reached an open space. The swimming motions he would use would be identical to what a human swimmer uses as he moves his body through water. However, in the case of a subterrestrial swimming through solid rock, the atoms of the rock pass directly *through* his body while, in the case of a human swimmer, the atoms of the water molecules flow *around* his body without entering it (unless, of course, he accidentally swallows some water!).

Once through the intervening rock wall, the subterrestrial being would then stop emitting the B rich anti-mass field radiation and his body would quickly resume its normal biological state of activity. As he finally emerged completely from the rock material, none of its various atoms would be left in his body. If the air temperature of the cavern he entered was excessively high, however, he might continue to emit the E rich anti-mass field radiation from his circulatory system that makes him immune to the potentially injurious tissue heating caused by the adsorption of the higher frequency infrared radiation given off by the surrounding atmosphere and rock walls.

Perhaps after these deepest dwelling subterrestrials colonize a certain area, they then artificially widen various natural passageways and even excavate new passageways so that they can more quickly move about without having to rely upon the use of their natural ability to interpenetrate solid rock. Such subterranean constructions would help to minimize the possibility of one of them accidentally getting "stuck" in solid rock and quickly dying should he tire while exerting himself and thereby not be able to voluntarily emit the B rich anti-mass field radiation at the rate necessary to maintain the interpenetration as his body traveled through the rock material.

With all of these remarkable abilities, one might assume that subterrestrial beings are also radically different from Earth's surface dwelling animals in other ways. I, however, am not convinced that this is the case. It must be remembered that, so far in this chapter, we have been describing beings who are remote ancestors of *human* beings! They are basically primates and must still eat food, drink water, breathe air, and reproduce just like any other mammal. They can also be expected to have some sort of family and community structures and to communicate with each other through spoken and written languages or via mental telepathy. The elaborate networks of tunnels that interconnect their caverns will probably be carefully marked with symbols for the purposes of identification and direction.

They will probably live in separated colonies at various depths that each consist of a cluster of interconnected neighboring caverns. Each colony will only have extensive interactions with other colonies near it whose members are similar in appearance and paranormal abilities. Like surface humanity, they will have their philosophies, religions, wars, and histories. Some of them will appear to be quite human in appearance while others will possess rather grotesque distortions of normal surface human features. I suspect that our planet's subterrestrials, with a few notable exceptions (like the "big foot" type creatures) will tend to appear to be more human-like as they inhabit underworld realms closer to the Earth's surface and will tend to appear less human-like as they dwell deeper within our planet's crust which will have required them to undergo much genetic mutation in order to survive.

There will, however, be some rather obvious differences between surface humans and our planet's subterrestrials due to the fact that they spend most of their lives living underground within the limited confines of natural and artificial rock chambers.

First and foremost of all, they will not be able to use fire. Remember, our remote ancestors had descended into the Earth's crust in order to escape the frigid surface temperatures brought on by the Ice Ages that were present during early human evolution. At that time, they did not yet possess the ability to make fire and, once inside the warmer interior of the Earth's crust, had no need for its use to keep warm. Also, the use of fire inside of their subterranean caverns and passageways is precluded by the limited air supply available to them there. Use of fire would have rapidly consumed this limited air supply and contaminated it with toxic gases like carbon dioxide and carbon monoxide.

Without the ability to use fire, these early subterrestrial beings would never have been able to develop any type of technology beyond the simplest of hand tools and implements. Every important survival ability that they would need, they would have to slowly evolve over the course of hundreds of thousands of years. However, once these traits had evolved they would be then be available for immediate use and not be subject to the many malfunctions to which artificial systems are prone. Thus, to levitate, become invisible, or interpenetrate solid rock, a subterrestrial being that had inherited these evolved abilities would merely have to exert his will to use them. He would not have to be burdened with carrying and maintaining any kind of technological equipment to generate these awesome effects.

One might wonder how our early subterrestrial ancestors managed to see their way along deep passageways if they could not use fire and had not yet evolved the ability to see in the infrared region of the electromagnetic spectrum.

To resolve this matter, we can imagine that, as these beings began to descend into the crust, they discovered other life forms already well established there. Animal life in the form of small fish and amphibians might have been found in subterranean lakes and pools. And, to complete the ecosystem, there would have to have also been plant life present. This plant life would have to consist entirely of molds, fungi, and higher forms that can produce edible vegetation without the need to use of photosynthesis. Perhaps the higher forms of subterranean plant life use "thermosynthesis" rather than photosynthesis; that is, they are able to convert the atmospheric humidity and subterranean animal life generated carbon dioxide gas into edible carbohydrates by using the thermal energy of the warm rock walls instead of the radiant energy of sunlight as do Earth's surface flora.

Since many species of surface molds and fungi are bioluminescent, it is not too much of a stretch of the imagination to assume that the plant life our subterranean ancestors encountered would have provided enough "cold chemical light" in the visible region of the electromagnetic spectrum for them to dimly see each other, move about, and explore their new environment. The ecosystem of animals and plants would have provided an abundant supply of food for these earliest of subterrestrials. As early hominid crust dwellers died off, their bodies would return organic materials to the ecosystem to maintain its equilibrium. However, expansion of a subterranean ecosystem would only be possible if fresh organic material could be brought down into the realms of the underworld. It can only be surmised that with the retreat and then reappearance of each new Ice Age, some new surface dwellers would be driven underground for shelter and would add to the biomass of the ecosystem already well established there.

With each new Ice Age, newcomers would arrive in the upper levels of the crust and have to deal with those more highly evolved subterrestrials already inhabiting caverns near the Earth's surface. There must have been much strife until, in time, the newcomers learned to communicate and live in peace with those who had arrived, perhaps, hundreds of thousands of years earlier. We can also imagine that with each new Ice Age and arrival of surface humans seeking shelter, the deepest dwelling subterrestrials would be driven even deeper into the Earth's crust and would, consequently, need to continue to evolve new paranormal traits for survival.

If the above scenario is accurate, then it is quite possible that right now the Earth's crust could be supporting a subterranean population that numbers in the millions! Yet, most of present day surface humanity is unaware of this startling possibility.

For the most part, I believe that Earth's subterrestrial population is quite content to remain separate from surface humanity. They, most likely, are curious about the surface world they once inhabited eons ago, but fear encounters with surface humans who may possess offensive technology that can be harmful to them such as firearms.

Of the many sighting cases which appear to involve subterrestrial beings that I have studied, it appears to me that they definitely have great interest in the products of surface technology that we humans possess which they can not make for themselves because they can not use fire. For example, those subterrestrials nearest the surface of Earth's crust may occasionally exit their underworld realms via various "portals". These underground passageways come so near the surface that it is possible for these beings to exit them through narrow openings or by actually interpenetrating only a foot or so of solid rock.

Once on the Earth's surface they regain their normal mass and weight by voluntarily *not* emitting anti-mass field radiation of any character and are then able to enter empty houses or stores in search of items to steal. They may take clothing, simple metal tools, grooming items, etc. These are mainly small objects which they can easily carry back to their underworld realms and which are then traded amongst the beings there. In the event of surprise discovery by a surface human, such a being can instantly reemit anti-mass field radiations of various characters from its body so as to turn invisible and escape through the walls of a buildings rooms. Any pilfered items carried close to its body will also share this invisibility and interpenetrability so that the items can be carried along with it.

Perhaps some day soon it may be possible to communicate with those subterrestrials who dwell nearest the Earth's surface. This will, no doubt, be extremely difficult to do and would only be possible if these timid creatures can feel absolutely sure that they will not be harmed or interfered with in any way. Perhaps then we will be able to get an accurate idea of their histories and what their daily lives are like. In exchange for this fascinating information we might supply them with some simple products of surface technology that could make their lives more comfortable and secure.

Of the subterrestrials dwelling closest to the Earth's surface, we must now briefly consider an elusive creature known as "Bigfoot" which is also called a "Sasquatch" or "Skunk Ape".

In the twentieth century there have been many reported sightings of these large, hairy, ape-like animals in the northwestern part of the United States. Indeed, reports of them are now surfacing in practically all states which have heavily forested regions. Even when they have not been directly observed, they have often left their deeply impressed footprints behind for people to discover. These prints can be large with a tip of big toe to back of heel measurement in excess of 16 inches (from whence the name "Bigfoot" was derived in 1958 by a news reporter) and, from the depths of their impressions in the ground, indicate body weights usually far in excess of two hundred and fifty pounds.

I became a believer in the existence of such creatures after viewing the famous Roger Paterson film that was made back in 1967 when

Paterson, a former rodeo rider, had been traveling on horseback with a friend through a forest in Bluff Creek which is in the northern part of the state of California. The 24 feet of film Paterson managed to obtain that day show a large, dark furred, ape-like being gingerly striding along on two legs near the edge of a stream bed at a distance of about 130 feet from Paterson. It has been estimated from the film and the footprints it left behind that the creature filmed was about 6 feet in height and weighed approximately 280 pounds.

What convinced me that this film was 100% genuine was a segment of it wherein the creature, while still continuing to move away from the camera, slowly turns and briefly looks directly toward the camera. Studying this movement in slow motion is most revealing. As the sunlight gleams off of the creature's shiny dark fur, one can make out the shape of muscles flexing under the fur in the legs. At chest level, pendulous breasts are momentarily visible which indicate that the subject of the film was a female! I also noted that when this Bigfoot turns toward the camera, she does not just turn her head, but rather swivels her entire upper body. This is movement that a large primate, like a great ape, must perform because the shortness of his neck makes turning just his head difficult to do without his large lower jaw hitting into his shoulder. Thus, he must turn his head by twisting his entire upper torso toward the direction in which he wishes to look. Such details, in my opinion, make it extremely doubtful that this film is a hoax.

One wonders exactly where such creatures go when they are out of sight and when night falls. Unlike such animals as apes, Bigfoots probably take shelter in easily accessible caves or subterranean caverns. However, because of its huge size and need for large quantities of calories and protein, it can not solely survive on the small amounts of food provided by a subterrestrial ecosystem. Rather, it is forced to leave its near surface subterrestrial environment and seek vegetation and small animals such as fish on the surface of our planet. Because of the powerfully strong body of a Bigfoot, it is probably safe from any other animal in a forest, including such ones as bears and mountain lions. It is, however, very vulnerable to human beings and their firearms. It is most likely for this reason that these creatures generally avoid humans.

If so, then it seems that they must have a way of communicating with each other and have used it to pass along stories of incidents in which they have been shot at, injured, or even killed by firearm carrying human hunters. There are tape recordings in existence that seem to indicate the Bigfoots use what has been dubbed "Samurai chatter" to communicate with each other. The words in this language sound somewhat like those in the Japanese language. And, other than the loud, shrill wailing sounds they can make, they can also make various grunting or "monkey-like" sounds.

In pondering the origin of Bigfoot type creatures, I concluded that they are nothing more then various species of evolved subterranean primates that belong to the genus known as "Gigantopithecantropus". Fossilized skeletal remains of this very distant cousin of present day human beings were found in China in the 20th century and it was assumed that they had been extinct for about a hundred thousand years. It now seems far more likely that the last Ice Age merely drove them underground to seek the warmth and safety of the Earth's crust. Because of their high metabolic needs, they must remain near the surface so that they can access the more abundant food supply there.

With modern technology, it might eventually be possible to capture a Bigfoot *alive*. If this could be done, then it might even be possible to develop some sort of simple sign language to communicate with it. This would allow for intelligent communication between humans and these creatures until their verbal language can be understood in detail. Possibly, this would allow us to get a general idea of how their tribes are structured. Then we could release the creature where it had been captured so that it could return to its own kind and tell of its experience with humans. However, to gain the trust of these very shy beings, we might let the one we set free take back a small supply of items with him that might be of use to such a large bipedal primate as it forages for forest food. Perhaps a small supply of metal axes, knives, and saws could be supplied to him. For the females among them, a supply of shatter resistant plastic mirrors that would allow them to view themselves without having to look into the still surface of a pond would be appreciated. With luck, the captured creature we freed would then later return to a designated meeting place for more items and we would be able to obtain more information from him as we continued to build up a bond of trust with the tribe.

If regular communication becomes possible with Bigfoots in the future, then it might even open up the possibility of surface humanity eventually communicating with *other* deeper dwelling subterrestrials. Again, trading surface manufactured items for information, it might be possible, in some future century, to unite the terrestrial and subterrestrial worlds of Earth on some level. One can only wonder what marvels will be revealed about the Earth and its past history if we are successful.

Let me conclude this chapter by briefly recounting what I believe to be a genuine subterrestrial sighting that occurred geographically close to me.

Shortly after I moved to my current residence in suburban New Jersey in the spring of 1991, I made friends with a neighbor whose large property was adjacent to mine. This neighbor had a daughter and her children living with him and one of the children was a young boy whom I shall refer to by the pseudonym of "Mikey".

During the spring of 1995 I was in our yard tending a small garden when I noticed that an advertising blimp was overflying my position at an unusually low altitude of only a few hundred feet. As this propeller driven craft slowly moved overhead, I heard Mikey over in my neighbor's yard let out a loud yell of excitement. He was about 12 years old at the time and had apparently never seen a blimp before at such close range.

He then popped his head up over the short stretch of stockade fence that borders part of his grandfather's property and asked me if it had been one of those flying saucers or UFOs that he had seen mentioned on a television program.

I explained to him that it was not, but rather was just a small blimp used for advertising and that there was nothing to worry about. This calmed him down and we then began to discuss the topic of UFOs and extraterrestrial life whose existence, despite his youth, he firmly believed in. I then asked him if he had ever seen a real UFO or anything else that was strange in his life. I have a habit of asking this question of people and have, over the decades, heard some very strange tales. But what Mikey began to tell me really caught my interest.

He told me that a few years prior to our conversation (at which time he would have been 10 years old and the year would have been 1993), he had been playing in his grandfather's yard toward late afternoon and decided to go to the extreme back part of the yard which is bordered by seldom used railroad tracks and a small creek. In order to reach the rearmost part of the yard, he had to walk around a large 24 foot diameter, above ground swimming pool which, because he was rather short for a 10 year old, completely blocked his view of the back of the yard.

As he moved around the side of the pool, Mikey suddenly became aware of a strange shape on the ground near the back of the yard. This shape was that of the body of a small humanoid creature that was squatting on the ground and holding something in its hands. This creature had greenish colored skin that appeared to be made up of small scales similar to that of a lizard's skin. Its head was small and on its forehead were some sort of comb like growths that projected away from the forehead. Although this creature never stood erect during this sighting, I believed that its actual erect height would have been about 4 feet.

Mikey could see that this being was eating a small bunch of edible wild berries that grew on several bushes at the back of the yard at that time of year (note: one should never consume any wild berries that one finds unless he is absolutely certain that they are *not* toxic). On each of his wrists, the creature wore a small gold bracelet! Aside from these accessories, this creature wore no other clothing.

As Mikey appeared from around the side of the above ground pool, the creature stopped eating the berries and slowly raised its head and turned it to stare at the child. Its eyes consisted of two large black pupils and, as its gaze fixed upon Mikey, he felt a wave of intense fear rising within him. The child then spun around and ran back to his grandfather's house as quickly as his short legs could carry him. Minutes later, some adults accompanied Mikey back to the spot where he had seen the small being, but by that time it had gone.

Needless to say, I found Mikey's story very interesting. I am inclined to accept his description of the alleged encounter as true and accurate because I knew him for about a decade during which time I had never found him to tell a lie. I had inquired about the incident from several of the adults in his family and at least one verified that he remembered the experience, but had dismissed it as the product of youthful imagination (in other words, it was thought that Mikey had made it up even though he had insisted that he had not). I consider it quite possible that Mikey is one of the very few humans who have ever been able to directly view a subterrestrial being.

Perhaps this creature had made its way to the surface and emerged from a subterranean crevice by interpenetrating solid rock to find itself standing with restored normal bodily mass and weight under one of the several small bridges that allow local automobile traffic to cross the small creek that meanders through our town. After a refreshing drink from the creek, this being was probably encouraged by the setting sun and dropping light levels to explore our surface world for a while. Making his way across the railway tracks, he soon found himself in the back of my neighbor's large property where he then discovered the berry bushes and their abundant supply of tasty berries. Not seeing anybody around, the creature decided to linger a while and sample some of its tasty treats.

Mikey's description of this small being is entirely consistent with what one might expect for a "typical" subterrestrial creature. It had a thick, scaly, greenish skin which, earlier in this chapter, was suggested would be evolved by subterrestrials in order to protect themselves from the higher radiation levels they would encounter as they explored and inhabited deeper levels within the Earth's crust. The greenish skin color could be due to a high copper content. The presence of this metal in sufficient concentration can effectively stop environmental radiation from reaching and damaging a subterrestrial's internal tissues.

The creature's large black eyes would have been mostly pupil and evolved in order to allow it to view its environment even in the dim light levels that are provided by subterranean bioluminescent vegetation or exclusively via infrared radiation emissions from warm rock walls. Mikey's creature most likely was also capable of telepathic communication.

Since many of the details of the mechanism responsible for telepathic communication were already treated in an earlier chapter, let me just briefly summarize the process here.

My past research indicated that telepathy is a natural means of communication that allows the transfer of mental images, feelings, and even sounds to occur between two minds without any visible means of transmission being involved. The process takes place when the mental images associated with conscious thoughts in the sender's mind are nearly instantly translated into a complex pattern of subconscious tremors in the rectus muscles that are normally responsible for moving the sender's eyeballs about in their sockets. These tremors, if unrestrained, would try to rotate the sender's two eyeballs about so as to cause their imaginary projected optical axes to trace out the shapes of the *edges* of the objects in the mental images on some imaginary plane surface placed in front of the sender's face.

During telepathic communication, however, these rectus muscle tremors are suppressed by the simultaneous contractions of each rectus muscle's *opposing* rectus muscle. This counter balancing effect prevents any gross rotation of the sender's eyeballs to take place but does cause *each* of the rectus muscles in the opposed pair that are "recruited" for the tracing process when a particular mental image enters his mind to emit a brief burst or train of extreme low frequency electromagnetic radiation or "ELF" of a fixed frequency and duration which then travels at light velocity to the receiver's eyes. There the incoming ELF trains are gathered and focused onto the receiver's corresponding set of rectus muscles. The receiver's rectus muscles then convert these incoming ELF trains into nerve impulses that pass via his optic nerves into the receiver's visual cortex where they eventually cause the sender's mental images to be formed in the receiver's conscious mind. Since this process works best with actual conscious mental images or with the unconscious images associated with conscious subvocalized words, it allows communication to take place between beings that, ordinarily, might not even *speak* the same language! (Note: For a far more thorough treatment of mental telepathy, the reader is referred to the author's book, *The Physics of the Paranormal*.)

Mikey's sudden appearance from around the side of the curving wall of the above ground swimming pool must have startled the subterrestrial being that was in the middle of enjoying his snack. At that instant various unconscious images in the creature's mind that were associated with his fear were almost instantly telepathically transmitted to Mikey's brain via a complex set of ELF trains as the creature turned his gaze on the child. The images would have appeared in the child's unconscious mind and, rather than be formed into actual images in his conscious mind, just resulted in a generalized panic reaction taking place in his nervous system. This accounts for the sudden wave of fear that overwhelmed Mikey and caused him to flee the scene.

The small gold bracelets that adorned the creature's wrists are also consistent with a subterrestrial origin. Gold is one of the few metals that occur in what is called a "native" state which simply means that one can find veins and nuggets of the metal in its pure elemental form (however, most native gold can contain up to 70% by weight of other metals such as silver, platinum, etc.). Since the subterrestrials can not use fire to smelt metals, they would naturally, like most primitive surface dwelling human cultures, tend to use native metals to fabricate bodily adornments. The bracelets Mikey's creature wore would have been made by hammering nuggets or short veins of the malleable gold into a long shape which would then be worn by simply wrapping the metal piece around a wrist several times. We can imagine that gold is far more available to Earth's subterrestrials than it is to us, but, like we surface dwelling terrestrials, they also value it for its beauty and durability.

Finally, I must analyze the remaining feature of Mikey's very rare sighting that contributes to its strangeness. This has to do with the bizarre comb-like growths that issued from the being's forehead. From their location, I believe that they may have had to do with the creature's sense of smell or even hearing. Mikey did not describe any sort of nose or ears on his creature. Possibly these functions became, through evolution, fused into a single centrally located organ on the creature's forehead. If these growths were for smell, then their position roughly corresponds to the location found on the human face. If they were for hearing, then their central location might also make sense. Humans have two ears so that they can locate the direction that sounds are coming from. In tight subterranean passageways, however, sounds are channeled by the rock walls so that they only travel along the length of a passageway. Under these conditions, two ears become unnecessary and a single, centrally located ear becomes sufficient to both hear and determine the direction of sound sources.

We see from cases such as the sighting by young Mikey that the subterrestrial inhabitants of our planet, on occasion, can and do ascend to our surface world. While fearful of surface humanity, they are curious about our existence and, perhaps, envious of the abundant food supply we enjoy and the fire based technologies we can utilize to make our lives more comfortable. For the present time, they restrict their explorations to the less populated regions of our surface civilizations and tend to emerge in the late evening or at night when the light levels are more comfortable for them. Possibly, their greatest fear is that we will discover their underground passageways and caverns and either attempt to force our presence upon them or even, out of irrational fear, try to poison their realms in an effort to eliminate these generally peaceful and benign forms of life.

Hopefully, surface humanity will allow future contact with these creatures to proceed at a pace that will make them feel as comfortable and unthreatened as possible. If that can be done, then the mutual benefits that might arise from future peaceful contact with these inhabitants of Earth's upper crust could be priceless.

Chapter 4

How Houses Get Haunted

THE SUBJECT OF HOW A dwelling can become a center for the many bizarre paranormal effects which cause us to declare it to be "haunted" can be a complicated one. For there are, literally, several dozen different effects that can appear in such sites and a thorough treatment of them is not possible here. However, in this brief chapter I will attempt to give the general reader interested in the subject a broad overview of these effects and must again refer those seeking a far more in-depth analysis to my book on the topic titled *The Physics of the Paranormal*.

I can start by stating that truly haunted houses are extremely rare and occur, in my opinion, far less frequently than genuine UFO sightings. However, unlike UFO sightings, all of which tend to be single, unique events, the phenomena displayed in real haunted houses have a tendency to occasionally repeat themselves on a somewhat regular basis. This attribute makes it just barely possible to apply the scientific method to these effects by collecting data on them via observations and instrumental measurements which can then serve in the formulation of working hypotheses that rationalize them.

Currently, worldwide, there are probably thousands of freelance or university affiliated paranormal investigators who are slowly accumulating the data needed to develop a holistic theory for the various phenomena encountered in haunted dwellings. Progress in the field is, unfortunately, hampered by two major factors. The first factor, of course, is the frustratingly capricious nature of the effects that have been noted over the centuries. The second factor is the problem of financing site research at locations where the effects seem to be manifesting with sufficient frequency so that meaningful observations can be made.

From my own study of this fascinating area of human inquiry, I have reached some definite conclusions concerning it. Mainly, I believe that the various phenomena that appear in hauntings can be roughly divided into five distinct categories in order of *decreasing* frequency of occurrence:

1.) The first and most frequently encountered category contains paranormal effects which are caused by currently living human beings who either reside in or visit the site of the haunting.

2.) The second category consists of the less encountered effects that are due to subtle changes in the structural materials of the dwelling that were caused by its *past* (and usually deceased) human residents or visitors.

3.) The third category includes the paranormal manifestations that are due to currently living *non*human beings who visit the site.

4.) The fourth category consists of seemingly paranormal effects which are, in fact, due to currently living human beings who are perpetrating hoaxes on paranormal investigators. This category would include any effects manifesting in the allegedly haunted dwelling which are being faked by its residents for some purpose such as publicity or financial gain. Because of this, the serious paranormal researcher must be ever vigilant that he is not being duped by witnesses who are hoaxing any effects being observed and / or fabricating the "testimony" being given so that the alleged events can be made to seem more credible because "independent investigators" have confirmed their reality "scientifically".

5.) The fifth and least frequently encountered category contains the phenomena associated with a past deceased human resident of the dwelling who has, through an extremely rare process, come back to life and now visits the former residence he enjoyed during his lifetime! As incredible as this would be, the existence of this rather bizarre fifth category seems, to me, to just barely be supported by the literature of the subject so I have decided to include it in this chapter.

There is, however, one cause for the various paranormal phenomena that manifest in haunted dwellings that I have not, to date, been able to find. After many years of studying this subject, I remain unconvinced that any of the paranormal phenomena associated with haunted dwellings might be caused by the return of the souls of human beings following their deaths. I know that this statement may surprise and possibly disappoint most readers, but the simple truth is that I, personally, have not seen anything in the "credible" research into the paranormal that convinces me yet that the effects experienced by witnesses actually involve "earthbound" human souls who are haunting a particular site because they are not yet "free" to "pass over" to the "other side". Thus, I do not consider paranormal phenomena as proof of the existence of an afterlife. This, however, does not mean that such a state of existence does not exist, but, rather, that I just do not consider the available evidence sufficient to prove its existence. For this reason there will be no mention of such things as souls or the hereafter in this chapter whose focus will be solely on the *physical* processes that may be occurring in haunted dwellings when paranormal effects manifest in them.

Let us begin the analysis with the first category of observed paranormal effects which, as was previously noted, are caused by currently living human residents of or visitors to a dwelling where the effects are being observed. Hereafter, I will refer to this category as "Category 1".

In this category one encounters a phenomenon known as the "poltergeist" or "noisy ghost". Often, in cases in which this phenomenon is manifesting, furniture can be heard sliding about in unoccupied rooms adjacent to the one containing witnesses and there may also be the sounds of glass breaking (all of these effects can also occur in the *same* room as witnesses occupy in which case they can be rather frightening). As soon as the noises start, they often quickly stop and surprised witnesses (who usually have only heard the sounds) will discover that, indeed, furniture in an adjoining room has been moved about or even overturned. The floor of the room may be littered with shards from broken bottles, vases, or dishes that mysteriously flew off their normal resting places to smash against walls or the floor. Books may have been pulled off of shelves and hung pictures torn off of walls. Several additional bizarre effects may also occur. These include the appearance of mysterious puddles of water or the far more dangerous outbreaks of fire in the flammable materials of furniture and drapery. Witnesses will often ascribe these effects to the actions of invisible earthbound souls who are trying to convey some message to the residents of or visitors to the dwelling.

In the vast majority of Category 1 cases, it is really one of the human witnesses present at the time of their occurrence that is responsible for the paranormal effects being manifested. I am not suggesting a deliberate hoax on their part, but, rather, that this person is unconsciously causing the effects. It has been noticed by several researchers that there seems to be a correlation between the incidents of poltergeist activity and the presence of human females in the dwelling who are either menstruating or are in their premenstrual phase at the time (however, I would definitely not exclude the possibility of these effects also being caused by a male). During these intervals of the female menstrual cycle there can be rapid swings in blood hormone levels that can then trigger the brain to cause the sudden vasoconstriction of blood vessels in various parts of the body.

As I suggested when treating this topic in my previously mentioned book, it may be possible for the iron containing hemoglobin molecules in human red blood cells to, under certain conditions, act like tiny magnets and to then align their magnetic fields with each other so that the human circulatory system can quickly become topologically equivalent to a large magnetic loop or circuit whose magnetic field moves rapidly along in the directions of its magnetic field lines inside of the blood vessels. Thus, the human circulatory system can, for a short time in this state, contain a strong magnetic "current" made up of many smaller parallel magnetic currents that "flow" through the circulatory system's various branches. The flow of these magnetic currents in a person's body is, like a person's normal blood flow, maintained by the pumping action of the heart.

As a female's blood supply rapidly flows through her vasoconstricted blood vessels, some of the electrons on the surface membranes of the blood cells will be deposited on the vessel walls and an electric field will build up between the blood cells and the vessel walls. This can be represented by electric field lines that connect the inside surfaces of the vessels with the surfaces of the moving blood cells. At this point a unique condition is created. One then has a magnetic field moving rapidly along the direction of its magnetic field lines which are everywhere inside of the vessels oriented *perpendicularly* or simply at right angles to the electric field lines which are also present.

Some readers may remember from my research into UFO propulsion that I have hypothesized that this process leads to the emission of "anti-mass field radiation". In the case of UFO propulsion, the anti-mass field radiation emitted from the craft's internal propulsion equipment acts to neutralize or cancel out the normally present "mass field radiation" that is emitted from the mass possessing subatomic particles that compose the various atoms of the craft and its crew. Since it is the mass field radiation from objects that gives them their gravitational and inertial properties, using artificially generated anti-mass field radiation to neutralize the craft and crew's mass field radiation emission will render them massless and, consequently, both weightless and inertialess. It is this state that then allows airborne UFOs to display the amazing performance characteristics noted for them while their crew members remain immune to the destructive inertial forces present during violent accelerations.

The reader should realize that the anti-mass field radiation and mass field radiation being discussed here are not the same as the photons found in the spectrum of electromagnetic radiations although they do travel with the same velocity as these which is about 186,000 miles per second. Rather, *both* anti-mass and mass field radiations are a form of *non*-electromagnetic *particle* radiation which is completely invisible to the human eye.

Indeed, I have found that much of the phenomena of the paranormal can be readily understood by assuming that anti-mass field radiation is strongly emitted from some people's circulatory systems in certain unique situations. When such anti-mass field radiation is produced, it can cause some rather mystifying local effects in the same room as witnesses or, when sufficiently intense, readily penetrate surrounding structures such as floors, walls, and ceilings to produce effects in adjoining rooms.

(I might briefly mention here that my research indicates that both anti-mass and mass field radiation emissions are composed of streams of elusive particles known as "antigravitons" and "gravitons". Both particles are massless, have an intrinsic "spin", and travel at light velocity away from their sources. Furthermore, these subatomic particles, when emitted in huge quantities, do not travel in isolation from each other, but, rather, assemble themselves into long "streams" of particles that are held together by weak coupling forces that act between them. If one could actually see these incredibly small streams of subatomic particles, each of which is far smaller than an electron, they might look like strings of pearls being fed out rapidly from their sources and extending off into space in all directions. More details of their structures and how they interact with each other can be found in my book *Secrets of UFO Technology*.)

Usually such anti-mass field radiation emission cannot be consciously controlled by an individual, but in the case of a human witness whose emission of it is responsible for the poltergeist activity in haunted houses, apparently its emission and the direction of that emission are under the *unconscious* control of that person. In other words, the unconscious mind of a certain very rare individual can actually focus and thereby intensify a beam of the anti-mass field radiation being produced by his entire body's circulatory system which will then produce local effects in the room he occupies or can actually penetrate some surrounding intervening structure of the room he occupies, such as a ceiling, wall, or floor, and then cause often quite noisy effects to take place in an adjacent room.

Sometimes, amazingly, if the person can only unconsciously radiate anti-mass field radiation in all directions about his body (that is, an *un*focused emission of this radiation), then he might actually begin to levitate off of the floor or out of a chair if the emission roughly matches the intensity and emission pattern of the normally produced mass field radiation emanating from all of the mass possessing subatomic particles contained in the atoms of his body and which is responsible for his normal bodily mass and weight. As his emission rate of anti-gravitons begins to approach his natural emission rate of gravitons, his bodily mass and weight will decrease. At some point he will be massless and therefore weightless enough to become buoyant and begin floating in the air.

In the previous chapter, we saw that anti-mass field radiation can further be considered to have three distinct "characters" depending upon how it is created and what effects it has upon the electrical properties of any matter with which it interacts.

The predominantly encountered type or character of anti-mass field radiation is the "neutral" character. It is used by airborne massless UFOs, levitating ufonauts and subterrestrials, and occasional levitating human beings in order to greatly reduce their weight and achieve atmospheric buoyancy and has no effect upon the electrostatic forces acting between electrically charged subatomic particles that compose the atoms of these objects. Because of this, neutral anti-mass field radiation only has the effect of negating or canceling out the *mass* field radiation naturally emitted by the mass possessing subatomic particles of objects which is responsible for their normal gravitational and inertial properties.

The second character was referred to as "E rich" and anti-mass field radiation of this type was able to weaken the electrostatic forces of *attraction* that normally exist between *unlike* electrically charged particles in matter such as exist between the electrons and nuclear protons within an atom. E rich anti-mass field radiation was involved in the process of rendering matter invisible in the optical or visible region of the electromagnetic spectrum.

The third character that anti-mass field radiation can have was referred to as "B rich" and anti-mass field radiation of this type was able to reduce the electrostatic forces of *repulsion* that normally exist between the *like* electrically charged particles of matter such as exist between the electrons within a polyelectronic atom or between the nuclear protons of two neighboring atoms. B rich anti-mass field radiation was involved in the process of rendering two separate pieces of matter capable of interpenetrating each other so that one piece could actually travel through the other without the various physical or chemical properties of either piece being affected.

If the natural mass field radiation emission of a UFO and its crew are completely negated by an artificially produced neutral anti-mass field of sufficient intensity, then they will become totally massless, weightless, and inertialess. In this state they can engage in the most violent of aerobatic maneuvers without the risk of damage to the vehicle or injury to its crew members. Apparently, a prolonged state of complete masslessness maintained by neutral anti-mass field radiation does no harm to a biological organism.

At this point it is also necessary to mention that there is another important feature of anti-mass field radiation emission which plays a very important role in the production of paranormal effects. This feature involves the "form" of the anti-mass field radiation or, rather, whether it is steady in its intensity or pulsates in its intensity as it is emitted at light velocity from its source.

If it is steady in intensity, then we can just describe the form as "steady". This is the form of neutral anti-mass field radiation that most levitating UFOs, unidentified flying ufonauts, subterrestrial beings, and occasional humans will produce and which simply negates the mass, weight, and inertial properties of the matter of its source, whether technological or biological, and any matter near that source. If, however, the anti-mass field radiation emission pulsates repeatedly over time, then it becomes necessary to talk about the *shape* of the pulsations of the "pulsating" anti-mass field radiation. Anti-mass field radiation pulsations of different shapes will affect any matter that they travel through in various ways and these will then determine what particular paranormal effects will be displayed by that matter.

To begin with, we can, as is done in the science of electronics, refer to the mathematical shape of any periodic *variation* in the *intensity* of emitted anti-mass field radiation with time as the "waveform" of that anti-mass field radiation.

The various types of levitations described above require that the waveform of the omnidirectional neutral anti-mass field radiation emitted from the levitating craft's or person's body have an emission intensity which remains relatively steady during the time that the radiation is emitted which means that there really are no separate pulses in it which are discernible. Thus, steady anti-mass field radiation of any character has no waveform associated with it. Because the *mass* field radiation normally emitted from all of the mass possessing subatomic particles of atoms is always steady in its intensity over time and the negation of mass field radiation is most efficiently accomplished by the use of steady neutral anti-mass field radiation, this is the type usually emitted by UFOs as they maintain a state of masslessness in either outer space or a planet's atmosphere.

My research indicates that, in the case of most poltergeist activity, the form of the anti-mass field radiation emitted by the causative person's body does *not* consist solely of the steady neutral type that would be the best for producing the complete levitation of his body as it most efficiently reduced his bodily weight to the point where he became buoyant. Rather, in the vast majority of poltergeist cases, only *some* of the anti-mass field radiation emitted from the causative person's body will be the steady neutral type, while the rest that is emitted will consist of a mixture of pulsating neutral anti-mass field radiations with differently shaped waveforms. Depending upon the shapes of the particular waveforms involved, the focused beams of neutral anti-mass field radiations emitted unconsciously from a causative person's body can produce different types of paranormal effects, either locally or remotely as in adjacent rooms if the beam is intense enough.

Next, we need to briefly discuss the various shapes or waveforms of *pulsating* neutral anti-mass field radiation and the effects they can produce on any matter as they pass through it at light velocity.

When a causative human's circulatory system emits pulsating neutral anti-mass field radiation, it usually mostly consists of pulses with a "sawtooth" waveform and, as the name implies, the shape of the intensity versus time graph of a series of these normally invisible and unnoticed pulses would look somewhat like the edge of common hand saw blade as they traveled away from the surface of the person's body at light velocity. One would notice that their peaks would be leaning and either sloping *away* from or sloping back *toward* the person who was emitting them.

But the human circulatory system can also produce pulses which have a waveform that is described as "sinusoidal". If one could view the intensity versus time graph of these invisible pulses as they traveled away from the surface of the causative human's body at light velocity, then he would note that their shape looked much like the symmetrical ripples that form on a pond when a stone is tossed into it.

At this point in this chapter the reader may feel a bit overwhelmed by the material that has so far been presented. Let us briefly summarize what has been discussed up to this point so that any confusion can be minimized.

We can start by stating that there are only two new forms of non-electromagnetic particle radiation which are being discussed here: mass field radiation which normally and continuously streams from the mass possessing subatomic particles of all of the atoms of any piece of matter and its opposite which is *anti*-mass field radiation that must be artificially produced by a UFO's propulsion system or can be naturally produced by a biological organism's circulatory system. It is the mutual cancellation of these two radiations when they are mixed together that is involved in the levitation of both UFOs and the occasional very rare human who can perform this feat.

While naturally produced mass field radiation is always steady in the intensity or rate of its emission, artificially produced anti-mass field radiation can be either steady in its rate of emission as happens when one uses it to produce the levitations of massive objects or it can be made to pulsate so as to produce a series of connected pulses of various shapes which are referred to as waveforms. The subatomic particles which compose both mass and anti-mass field radiations always travel away from their sources at light velocity. Furthermore, it is possible to cause anti-mass field radiation to be emitted from a source in the form of a beam and it is such beams, when emitted involuntarily from an *unsuspecting* human's body, that are, I am very firmly convinced, involved in the production of the various poltergeist effects.

Continuing the summary of what the reader has so far learned, it can be stated that anti-mass field radiation can further be described as having one of three "characters" or types depending upon the balance of the intensities of the electric and magnetic fields that were used to produce that radiation's component particles. Thus, anti-mass field radiation could be described as either "neutral", "E rich", or "B rich".

Neutral anti-mass field radiation was produced when the intensities of its generative electric and magnetic fields were balanced. E rich anti-mass field radiation was produced when the intensity of the electric field was stronger than that of the magnetic field. And, finally, B rich anti-mass field radiation was produced when the intensity of the magnetic field was stronger than that of the electric field. Neutral anti-mass field radiation was the type mostly produced by levitating UFOs and humans. E rich was involved when either a UFO or a subterrestrial being turned invisible and their atoms' electrons no longer interacted with photons of light in the optical region of the electromagnetic spectrum. B rich was involved when a subterrestrial being briefly interpenetrated a solid obstacle such as a rock wall separating two adjacent subterranean caverns. Since mass field radiation is produced naturally by all of the mass possessing subatomic particles in the atoms of matter, it is not yet known if it can also have various characters and we can probably safely assume, at this point in time, that all mass field radiation is always neutral in character.

In order to fully rationalize paranormal phenomena such as occur in haunted houses, however, it is not sufficient to have only three types of anti-mass field radiation possessing different characters. We learned that these three types of anti-mass field radiation are subject to a further subdivision based on their "form". Their form could be either steady or pulsating.

If the anti-mass field radiation's form was steady, then there was little if any variation in its intensity as it radiated out of its source at light velocity whether that source was a UFO's propulsion equipment or a subterrestrial's or human's circulatory system. If the form was pulsating, however, then we needed to further *subdivide* its form into three distinctly shaped "waveforms".

Thus, pulsating anti-mass field radiation (whether its character is neutral, E rich, or B rich) can be described as having a waveform that is either: (a) "sawtooth" with the peaks leaning away from the source producing it and out into the direction that the radiation travels at light velocity; (b) "sawtooth" with the peaks leaning back toward the source producing it and opposite to the outward direction that the radiation travels at light velocity; and, finally, (c) "sinusoidal" in which case the pulses are shaped like sine waves or like the ripples that travel across the surface of water. One can think of the sinusoidal waveform as consisting of a series of peaks that point up and neither lean away from or back toward their source of emission, but, rather, always point in a direction at right angles to the direction of travel of the anti-mass field radiation.

So, when we consider the *three* characters of anti-mass field radiation (neutral, E rich, and B rich) and the *two* forms these can *each* have (steady or pulsating) and the *three* waveforms shapes (peaks leaning away from the source, peaks leaning back toward the source, and sine waves whose peaks neither lean away or back toward the source) that *one* of those two forms (the pulsating form) can have, we have a total of *twelve* distinct types of anti-mass field radiation that are involved in the production of the various paranormal effects that occur in "haunted" houses!

Finally, to complete our brief summary, here is a list of the twelve different types of anti-mass field radiations possible. The first name in each type's listing is the character of the anti-mass field radiation. The second name is its form which will be either steady or pulsating. The third phrase or word in the listing, *if the radiation is pulsating*, will describe its waveform which will be either sawtooth with peaks leaning out away from the emission source, sawtooth with peaks leaning back toward the emission source, or sinusoidal. Quite fortunately, the most frequently observed of the Category 1 paranormal effects occurring in haunted houses can be adequately rationalized by using only the first four types of anti-mass field radiation on this list.

1.) neutral > steady
2.) neutral > pulsating > sawtooth peaks leaning out away from the emission source
3.) neutral > pulsating > sawtooth peaks leaning back toward the emission source
4.) neutral > pulsating > sinusoidal
5.) E rich > steady
6.) E rich > pulsating > sawtooth peaks leaning out away from the emission source
7.) E rich > pulsating > sawtooth peaks leaning back toward the emission source
8.) E rich > pulsating > sinusoidal
9.) B rich > steady
10.) B rich > pulsating > sawtooth peaks leaning out away from the emission source
11.) B rich > pulsating > sawtooth peaks leaning back toward the emission source
12.) B rich > pulsating > sinusoidal

With this level of complexity, is it any wonder that so little progress has been made in the last several hundred years in understanding the details of the physics of paranormal phenomena? However, as surface humanity eventually learns how to artificially generate anti-mass field radiation with its various characters and forms, I am very confident that we will very soon thereafter be able to artificially produce all of the paranormal phenomena that have been reported in the literature so far. I can only hope that my early researches into this topic will provide some guidance to future researchers as they tackle the details of this subject.

With the reader now having knowledge of the twelve types of anti-mass field radiation possible, let me provide a brief overview of how the various types of this radiation can produce the most frequently encountered Category 1 paranormal effects noted in haunted dwellings.

In the case of poltergeist activity that involves the generation of motive forces on furniture and small loose items that might be found on a table or shelf, it is actually a *combination* of both steady and pulsating neutral anti-mass field radiations being simultaneously emitted from the causative human being's body that accounts for this paranormal phenomenon. The pulsating neutral anti-mass field radiation portion will then have one of the two possible *sawtooth* shaped waveforms with the particular waveform involved determining which direction the resulting motive forces will act on the objects.

As an object, either in the same or an adjacent room to the causative human, is simultaneously penetrated by a focused beam containing both of these forms of anti-mass field radiation, the sawtooth pulses of the pulsating neutral anti-mass field radiation component will, as they pass through the object at light velocity, actually result in the formation of *stationary* crossed or perpendicular electric and magnetic fields *inside* of the object (note that these fields have steady intensities and exist only so long as the object is penetrated by the pulsating neutral anti-mass field radiation and they remain stationary with respect to the object's environment even if the object begins to move relative to that environment and the fields).

Exactly why these two orthogonal fields should form is still a bit of a mystery to me, but their presence obviously has something to do with the continuously changing intensity of the neutral anti-mass field radiation inside of the object as the radiation passes through the object. The situation seems somewhat similar to a phenomenon observed in electrical science in which an electric field changing at a fixed rate produces a magnetic field of constant intensity while a magnetic field changing at a fixed rate produces an electric field of constant intensity. Perhaps the rapid pulsations of the neutral anti-mass field radiation inside of the object somehow cause some of the antigraviton particles from which it is composed to breakdown and then form electric and magnetic fields at right angles to each other. This seems logical since antigravitons are originally created as a result of electric field lines interacting at right angles to magnetic field lines either artificially inside of a UFO's anti-mass field generator or naturally inside of a levitator's circulatory system. Obviously, determining the exact mechanism involved in the formation of the perpendicular electric and magnetic fields that are necessary for the application of "telekinetic" (sometimes referred to as "psychokinetic") forces to objects as happens during poltergeist events will be a matter for future physicists to work out.

Once the pulsating neutral anti-mass field radiation portion of an incoming beam from the body of a causative person establishes the stationary electric and magnetic fields inside of an object, the *magnetic* field component created can then readily interact with the other *steady* neutral anti-mass field radiation portion of the beam which is also penetrating the same space inside of the object. This interaction is then identical to that which occurs near the hull of a massless, airborne UFO when its propulsion system uses the interaction between its emitted steady, excess neutral anti-mass field radiation and the magnetic fields projected out beyond its hull to ionize the air surrounding its hull and thereby convert it into a rich plasma composed of various atmospheric ions. (The exact mechanics of this ionization process are a bit complicated and have been covered in my second book on UFOs, *The How and Why of UFOs*. Let it suffice here to just say that as steady neutral anti-mass field radiation passes through a magnetic field located outside of the source of the steady neutral anti-mass field radiation, some of that anti-mass field radiation is converted into a form of B rich anti-mass field radiation that, somewhat paradoxically, behaves much like steady E rich anti-mass field radiation which can then weaken the attractive electrostatic forces acting between *unlike* electrically charged subatomic particles such as exist between an atom's orbiting negatively charged electrons and its positively charged central nuclear protons. This then results in extensive ionization taking place as the atom's own thermal energy is used to extract the outermost loosened electrons from the atom. As the atoms and molecules of an atmospheric plasma continue to loose thermal energy as ionization proceeds and the concentration of "free" electrons in the plasma increases, the plasma's temperature will also continue to drop.)

In a similar manner, the interaction of the steady neutral anti-mass field radiation portion of the beam with the magnetic field temporarily created inside of the object causes an immediate weakening of the electrostatic forces of *attraction* that hold the electrons and nuclear protons of the atoms of the object together. As a result, the outer electrons in the atoms of the object's structure being affected by the projected beam from the causative person's body are more easily removed from their atoms even though the object might be fabricated from materials such as glass and wood that normally aren't electrically conductive as are metals. This then allows the ambient thermal energy of the object to actually highly ionize its *own* atoms and molecules. During this process, the object's temperature can suddenly drop many degrees depending on the material out of which it is composed.

The resulting ions that form inside of the object will then find themselves in a region of space that is also penetrated by crossed or perpendicular stationary electric and magnetic fields of steady intensities. The ions in the object will then "feel" what are known in physics as "Lorentz forces" acting on them that will cause them to move in the *same* direction regardless of whether they are negatively or positively electrically charged and to always move at right angles to *both* the electric and magnetic field lines. Thus, these ions will, depending on the orientation of the directions of the electric and magnetic field lines relative to each other inside of the object, *all* be either pushed away from the source of the incoming anti-mass field radiations (which in the case of poltergeist phenomena is a human's circulatory system) or pulled back toward it. Because the positively electrically charged ions are still part of the normal molecular structure of the object (most of whose atoms are not ionized), the entire object itself will feel a force acting on it that tries to move it.

If the force acting on an object is strong enough, it will begin to move. This motion will usually be a sliding one (facilitated by objects placed on polished surfaces), but may also be one that tilts or even lifts the object. The resulting direction of motion of the object is always either away from or toward the causative human source of the projected beam containing the *two* types of anti-mass field radiation that are passing through the object. The particular direction of motion of an object is, however, ultimately determined by the direction in which the time varying intensity peaks of the sawtooth pulses of the pulsating neutral anti-mass field radiation portion of the beam lean which then determines the orientation of the directions of the created electric and magnetic fields lines relative to each other inside of the object.

Thus, if the sawtooth pulses of pulsating neutral anti-mass field radiation passing through the object have peaks that lean away from their human source and toward the outward direction that the pulses travel, this might result in an electric field being created inside of the object whose field lines point *up* and a created magnetic field whose field lines will be oriented at right angles to that direction and therefore point to the *right* of the object as viewed from the location of the causative human (actually the electric field lines could point in any direction *perpendicular* to the path of the beam, but I've selected this particular direction here in order to make what's happening easier to visualize). This would then cause the newly formed ions inside of the object to feel Lorentz forces acting on them that push them *and* the object they occupy *away* from the human that is the source of the beam.

On the other hand, if the sawtooth pulses of pulsating neutral anti-mass field radiation have peaks that lean back toward their human source and away from the direction in which the pulses travel, this might result in the creation of an electric field inside of the object the direction of whose field lines still point *up*, but results in the creation of a magnetic field whose field lines now point to the *left* of the object as, again, viewed from the location of the causative human. This change in the orientation of the directions of the created electric and magnetic field lines with respect to each other then causes the newly formed ions inside of the object to feel Lorentz forces acting on them that pull them and the object they occupy *toward* the human who is the source of the two types of anti-mass field radiation passing through the object at light velocity.

We see from this that objects of various sizes and masses in adjacent rooms (or even the same room) will either move away from or toward the causative human's body during certain brief time intervals when his circulatory system is emitting the focused beam of the *two* required different types of anti-mass field radiations with sufficient intensities. It's important to note here that the force applied to the object is always along a line that, if extended, would lead directly back to the causative human's body. As a result, generally, the object's motion will almost always be along that line and *not directly* at right angles to this line which is just the projection path of the beam.

However, if the object is a cart on fixed wheels whose planes are fixed at an angle to the line of projection of the beam, then it is possible for the cart to move toward or away from the human source while also undergoing *some* lateral motion perpendicular to the projection path of the beam.

In this case, the fixed wheels force the cart to roll away from the projection path as the beam applies pushing or pulling forces to the cart. We can refer to this type of slantwise motion of objects as "tacking" after the term for the maneuver used in sailing in which a sailboat can actually head into the wind, but must do so by following a zigzagging course. Obviously, for the cart to move in this way any significant distance requires that the projected beam remains focused on the cart during its extended lateral motion. This requirement implies that there is probably some sort of feedback mechanism at work which, like radar, gives the causative human's subconscious mind information about the position of the cart even though it may be hidden from his view by an intervening wall and this information is then used to keep his projected beam "locked" on the target object as it moves.

So far I have assumed for the sake of discussion that the adjacent room was on the same floor as the causative human. This need not be the case and the situation becomes more complicated when the poltergeist effects occur in adjacent rooms that are either above or below the floor of the room containing the human source of the beam of the two types of anti-mass field radiation. In these cases, the motions of the objects producing the sounds as they fall over can be sliding ones or they can involve actual levitations of an object as it is physically lifted from its location and then suddenly allowed to crash to the floor when the lifting force applied to it is interrupted.

If the causative human only weakly emits the two required neutral anti-mass field radiations from his circulatory system, one steady and one pulsating in intensity, or poorly directs them (a process requiring the unconscious use of muscle fibers in the blood vessels to temporarily deform the vessels so that they emit most of their anti-mass field radiations in selected directions and can thereby focus them into a beam that will converge on a distant object), then only smaller and less massive objects such as books or dishes may take flight and these Category 1 effects may only occur in the same room occupied by the causative human since passage of the beam through a wall into an adjacent room would weaken it to the point where it could not affect even small objects.

In such cases, it usually eventually becomes obvious that the effects are associated with a particular single person in the dwelling. When that happens, it may be falsely assumed that the person is "haunted" or "possessed" by a malicious spirit. If, however, the person is a strong emitter of the two necessary types of anti-mass field radiation, then he or she may be able to slide pieces of furniture about in adjacent rooms about if they are light enough or mounted on casters, knock over free standing items such as book shelves or grandfather clocks, or cause hung paintings to be pulled off of walls.

So, the reader can see from this that all poltergeist activity that is produced by the circulatory system of a current human occupant of a haunted house is, in reality, a form of telekinesis or the moving of objects at a distance with no apparent physical means involved. By employing the concept of the projection of a focused beam consisting of two types of anti-mass field radiation passing through the target object, the effect becomes more understandable. As previously noted, this beam then has two distinct effects on the object. It induces ionization of the atoms *within* the object and then it causes the motion of the object by applying Lorentz forces to the ions that were formed.

The motion produced, however, is still subject to our other well known laws of physics, particularly Newton's Third Law of Motion. If a young female human is seated in one room and her projected focused bodily neutral anti-mass field radiations, one steady and the other pulsating, cause a heavy object to begin moving away from her in an adjacent room, one might wonder why her body does not react by being trust back in the opposite direct at the same time and, perhaps, knocking her out of her chair. The solution to this mystery is that her circulatory system actually simultaneously emits *two* projected beams of anti-mass field radiations that travel in *opposite* directions away from her circulatory system.

The main beam, discussed above, might penetrate a nearby wall and only apply weak forces to the wall because the beam is not focused and thereby intense enough at the spot where it passes through the wall. The wall, being securely attached to the floor and ceiling, undergoes no motion from the weak telekinetic forces applied to it. However, once inside the adjacent room (which could be on the same floor as her room or above or below that floor) the beam becomes focused enough and intense enough to apply a force to some object in the room that then causes that object to begin moving either away or toward her location. As this is happening, a *secondary* beam is simultaneously projected from her body in the exact opposite direction to that in which the main beam is projected.

That secondary beam then applies an *equal* force to some nearby stationary object in the opposite direction, such as one of her room's walls (or either its ceiling or floor if the adjacent room where the poltergeist activity is occurring is either below or above her room's floor). The result is that the two reactional forces produced on her body by the two oppositely projected anti-mass field radiation beams (which are both trying to *simultaneously* either push objects away from or pull them toward her) will always cancel each other out and she will not feel any forces being applied to her body. Even though a heavy object may be moving about in an adjacent room, she will always remain completely stationary and unaware of anything happening.

Let us now consider another classic Category 1 effect that as been observed and recorded from ancient times. I refer to the phenomenon of mysterious fires that can occur without warning in dwelling and these can also be attributed to the presence of a causative human who is capable of projecting a beam containing two types of anti-mass field radiation. As in the case of poltergeist or telekinetic effects, one of the components of the beam will be steady neutral anti-mass field radiation. However, the other component of the beam will be pulsating neutral anti-mass field radiation which has a *sinusoidal* rather than a sawtooth waveform.

Similar to what takes place when poltergeist effects manifest, mystery fires occur when a focused beam of steady and sinusoidally pulsating neutral anti-mass field radiations emitted from a causative human's circulatory system penetrate an object. The sinusoidally pulsating neutral anti-mass field radiation also immediately creates stationary electric and magnetic fields inside of the object. However, in the case of pulses with a sinusoidal waveform there is a big difference. We can imagine that *each* complete pulse of sinusoidal neutral anti-mass field radiation passing through an object acts like two "back to back" sawtooth pulses with different waveforms that alternately affect the object. The first half of the sinusoidal pulse is like a sawtooth pulse with its peak leaning back toward the human source while the second half of the sinusoidal pulse behaves like a sawtooth pulse with its peak leaning away from the human source.

The first half of the sinusoidal pulse might then cause the created electric field lines inside of the object to point up while the created magnetic field lines (again as viewed from the location of the human source) point to the left of the beam's path. Then the next half of the sinusoidal pulse will cause the created electric field lines to still point up while the created magnetic field lines suddenly disappear and then reform so that they point to the right of the incoming beam's path.

In this scenario, the magnetic field lines created inside of the object by the sinusoidally pulsing neutral anti-mass field radiation component of the beam actually rapidly and repeatedly reverse their direction with respect to the created electric field lines which remain stationary and always pointing up. This rapidly oscillating magnetic field will cause little if any gross motion in the heavier positively charged ions of the structural atoms of the material, but it will cause the much lighter negatively charged electrons extracted from those structural ions (with the energy for the ionization process again provided by the thermal energy of the material) to begin surging back and forth inside of the ionized material of the object along the path of the projected beam passing through it. (However, when the target object is metallic and therefore normally electrically conductive, the rapidly oscillating magnetic field inside of it may have an external field that is strong enough to temporarily "couple" with or be attracted to a nearby unaffected heavy metallic object or some other metallic structure in which case a less massive target metallic object may begin violently vibrating and, as a result, "dance" its way off of a table or shelf. The target metallic object need not be made of steel.)

Thus, the effect of a beam of steady and sinusoidally pulsating neutral anti-mass field radiation on a target object is to cause an alternating electrical current to be set up in it. As the ionization process lowers the electrical resistance of materials that are normally electrically *non*conductive such as cloth or paper, these materials will begin to act like high resistance conductors. The object will then, as long as the focused beam from the causative human's circulatory system is penetrating it, have a heavy electrical current surging back and forth from one side of it to the other (with the left to right side changes in the directions of the oscillating magnetic field lines inside of the object being used here, the induced current inside of the object would flow rapidly toward the causative human and then away from him and would repeat this action for every sinusoidal pulse passing through the object) and this alternating current will cause the material of the object to begin to heat up.

This temperature rise will occur despite any cooling that took place in the object as the ionization of its structural atoms took place. When its temperature reaches in excess of about 450 degrees Fahrenheit the object's material will ignite and the resulting fire can then quickly spread and set the entire room on fire that contains the object. In small towns and villages of the past that did not have adequate firefighting technology, this could easily result in the loss of an entire building.

As in the case of the focused beam that was described above that produced poltergeist effects, the beam emitted from the causative human's body also would be accompanied by a secondary beam in the exact opposite direction. That beam might also induce a fire in a suitable material in that secondary direction, but, then again, depending on the nature of the material it encountered (for example, if it was stonework of some sort) nothing might happen.

This effect has been noted repeatedly in the literature of the paranormal through the centuries and has often been associated with women who experience rages. Perhaps their rages and the resulting fires occurred only during certain phases of their menstrual cycles, but, as in the case of the focused beams that produce telekinetic effects, the changes in the causative person's circulatory system will also involve vasoconstriction of some sort.

There are many other Category 1 type paranormal effects that occur in haunted houses which can also be adequately explained by assuming the emission of focused beams of various types of anti-mass field radiation from a causative human being's circulatory system whether it be that of a male or female. These include such effects as unexplainable puddle formations in rooms, water spraying mysteriously from unbroken walls, small items from *outside* the dwelling being teleported into the room and then suddenly materializing and appearing to drop out of a ceiling, etc. Unfortunately, it is beyond the scope of this short chapter to treat all of them here in detail. Rationalizing them requires additional use of the other types of anti-mass field radiation listed above; that is, types #5 through #12. The serious student of this subject is again referred to my book, *The Physics of the Paranormal*, for a more extensive treatment of these effects.

Next, we must deal with the second category of paranormal effects seen, with far less frequency, inside of haunted dwellings that involve phenomena that are currently manifesting, but which are due to past and usually deceased occupants of the dwellings. As examples of these Category 2 effects, let us consider the cases wherein vague glowing forms are seen in older buildings and other cases in which contemporary "psychics" or "sensitives" are able to accurately detect the location and nature of the activities of past residents in dwellings.

Let's begin with the stereotypical image that springs to mind when we hear the word "ghost" mentioned. We usually imagine an indistinct, slightly glowing figure that either suddenly appears floating in mid-air in a room or which may appear and then slowly drift down a hallway or even up or down a flight of stairs.

When static, such a slightly luminous entity may just slowly fade away or, if moving, appear to drift away or even disappear by penetrating into a solid wall! Needless to say, people who witness such a manifestation find it to be rather frightening. However, sometimes the level of anxiety of witnesses will diminish when they realize that these apparitions have a tendency to reappear on a more or less regular schedule. Often the glowing figure will appear at around the same time of night, at about the same time of year, and, most consistently, in approximately the same location in the haunted dwelling. Fear is also further diminished when one realizes that these apparitions do not intelligently interact with witnesses or even seem aware of their presence.

The best current explanation for these Category 2 types of ghostly manifestations is that they represent a kind of holographic image that is formed in the air of a room or other space inside of a dwelling. The source pattern for this image was originally stored in the building materials of the dwelling's room or other space as a result of processes that were, at some past time, occurring inside of the body of a past resident or visitor to the dwelling. Then, at a later date which could be years, decades, or even centuries later, this source pattern is activated under certain precise conditions and results in a glowing and usually indistinct holographic three-dimensional image of the past resident forming in the air of the space enclosed by the floor, walls, and ceiling of the room whose materials contain the invisible source pattern. The image will be formed at approximately the same location (with respect to the space's enclosing surfaces) that was occupied by the past resident when he originally impressed the source pattern into the building materials surrounding his location in the room while he previously occupied it.

As was noted in the previous discussion of poltergeist activity, during times of physiological stress in which vasoconstriction occurs, the human circulatory system can emit two types of anti-mass field radiations one of which is steady in intensity and one whose intensity pulsates with variously shaped waveforms. One type of pulse with the so-called "sinusoidal" waveform was seen to be implicated in cases of mysterious house fires that appeared to start for no obvious reason. We saw that when a focused beam of such anti-mass field radiations left the activated circulatory system of the causative person and then reached and passed through nearby combustible materials, it could render normally nonconductive materials somewhat electrically conductive and would then induce strong electrical current flows in these materials that could heat them until ignition took place.

This mechanism can be readily modified to provide a theoretical basis for the Category 2 effects occasionally observed in haunted dwelling.

If a past living human resident of a dwelling experienced some sort of brief, but extreme physiological stress that resulted in him emitting both steady and sinusoidally pulsating neutral anti-mass field radiations from his circulatory system, then these radiations could easily reach and then pass through *all* of the floor, wall, and ceiling surfaces immediately surrounding him of the room in which he was located. In these cases, unlike what happens when Category 1 poltergeist telekinesis or mystery fires occur, the person's body does *not* emit two oppositely directed beams of the two types of anti-mass field radiations, but, rather, his bodily emission of these two types of anti-mass field radiation is omnidirectional and actually bathes the entire surrounding surfaces of the room he occupies.

At these surfaces, any *steady* anti-mass field radiation also emitted from the person's body would then combine with the temporary stationary magnetic field created in the surface materials by the sinusoidally pulsating neutral anti-mass field radiation to form B rich anti-mass field radiation that behaves like E rich anti-mass field radiation (the same kind created outside of the hull of an airborne massless UFO and previously mentioned) beneath the interior surfaces of the building's structural materials. With the atoms of the surface materials exposed to this type of E rich anti-mass field radiation, the result would be the immediate weakening of the *attractive* electrostatic forces acting between the atoms' subatomic particles having *unlike* electrical charges such as those which serve to keep negatively charged electrons orbiting their atom's positively charged nucleus. This weakening then causes ordinary building materials such as wood, plaster, stone, etc. that are usually electrically nonconductive to begin to behave like electrical conductors as a large supply of loosely bound conduction electrons begins to accumulate in them. These electrons will be the ones orbiting farthest from their particular atom's nucleus. At this point these materials actually begin to behave, electrically, as though they were metals!

Since these conduction electrons produced inside of the surrounding walls, ceiling, and floor are then exposed to pulsating neutral anti-mass field radiation created stationary electric and oscillating magnetic fields inside of these surfaces, the electrons will be forced to swirl about in various ways and, depending upon the location of the causative human and his bodily posture in a room, will form a very complex and unique pattern of eddy currents in the building materials which will remain stable as long as the room surfaces are exposed to the human source's two types of emitted anti-mass field radiations.

These electron eddy currents, however, also have a critically important secondary effect. They then force an actual physical change to take place in the microscopic crystalline structures found in the room's surface building materials. Possibly, this transformation of a building material's original crystal structures is facilitated by the removal of the outer electrons of its atoms that took place when they were ionized. With some of their outer electrons temporarily liberated from them, these atoms can, since they are in a constant state of vibration because of their thermal energy, then shift about and reorientate themselves more easily in response to the electron eddy currents so as to form a new pattern of crystal structures.

Thus, the invisible crystal structures inside of the various building materials that compose the interior surfaces of a room were originally randomly orientated when the materials were first put in place during the construction of the dwelling, but the temporarily formed pattern of electron eddy currents described above changes this situation and forces the atoms in these microscopic crystals to actually physically reorientate themselves so as to also form a new complex and unique holographic pattern within the materials of a room's interior surfaces. Once this new invisible pattern is fixed, it remains permanent. It's important to realize here that this change in crystal structure of the building materials that takes place is a very subtle one. It produces no visible changes to the naked eye, but might possibly be detected by some advanced method such as x-ray diffraction analysis.

Then, at some future time, environmental changes can take place in the previously altered construction materials that will then cause each of the impressed microscopic crystal structures in them to *temporarily* act like a microscopic anti-mass field generator and begin emitting steady neutral anti-mass field radiation. However, there will have to be a fairly precise set of conditions that must all exist simultaneously in order to cause this emission to take place.

I suspect that the flow of heat energy through the building materials and the resulting temperature differences between rooms and even between a dwelling's interior air and the ground under the dwelling have much to do with the activation process.

As the temperatures of these various materials increase, they will physically expand and the pressure from this expansion will, through something known as the piezoelectric effect in electronics, cause an increasing number of electron eddy currents to form among the materials' previously reorientated microscopic crystal structures. There could actually be millions or even billions of these temporarily stable eddy currents set up inside the interior structural surfaces of a single room! The resulting orientations in space of these many microscopic eddy currents will then be determined by the arrangement of the reorientated crystal structures nearest them. Most importantly, these *later* piezoelectrically induced electron eddy currents will exactly duplicate the original ones created in the dwelling' interior surface building materials by the two types of anti-mass field radiations which were provided by the circulatory system of some previous visitor or tenant of the dwelling. That past causative human resident might still be alive or he could have been dead for years, decades, or even centuries.

At this point, these piezoelectric eddy currents and the microscopic crystal structures which control their orientations in space will begin to act like microscopic anti-mass field generators and will emit steady neutral anti-mass field radiation which is then projected back into the interior of the dwelling.

Actually, the steady neutral anti-mass field radiation issuing from each previously altered microscopic crystal structure in the room's interior surface building materials and their newly associated eddy current will not be omnidirectional, but, rather, will be directed so that it travels back to the *general* location inside of the room which contained that portion of the past resident's stressed circulatory system that originally emitted the two types of anti-mass field radiation that resulted in the creation of that particular altered microscopic crystal structure. Thus, each portion of the space inside of the room which corresponds to the location of a particular portion of the past resident's circulatory system will be receiving steady neutral anti-mass field radiation from millions or even billions of microscopic reoriented crystal structures and their associated piezoelectric electron eddy currents which are distributed, more or less, uniformly over every square inch of the interior surfaces of the room.

In a very real sense it is as though a holographic pattern of the past human resident's circulatory system had been stored in the structural surfaces of the room he once occupied in the form of a holographic pattern of reoriented microscopic crystal structures. Once activated by a specific set of environmental factors, this stored pattern can then create a three-dimensional image of the person's circulatory system which will form in the air of the room at the location he previously occupied.

As happens in ordinary still photography, this later formed image within the interior space of the room, due to some bodily motion of the past human resident when this holographic source pattern was originally impressed into the dwelling's interior surface materials, will be slightly out of focus and, initially, *invisible* to the human eye! Once formed, this three-dimensional image inside of the room will be maintained only so long as the steady neutral anti-mass field radiation forming it converges on the various portions of the volume of space it occupies. This image, like many transient environmental phenomena, will probably exist for only about a minute or so and will quickly fade as soon as the specific environmental factors stimulating its formation no longer exist.

If a magnetic field is present in the room from some source such as a large nearby magnetized iron object, then the steady neutral anti-mass field radiations that are converging from the room's surrounding interior surfaces so as to form the blurry three-dimensional image of the past resident's circulatory system will combine with this magnetic field. And, again, this combination will result in the production of the special type of B rich anti-mass field radiation which behaves like the E rich anti-mass field radiation that causes the reduction of the electrostatic forces of attraction that act between the unlike electrically charged subatomic particles that hold individual electrons inside of their respective air molecules and atoms. When this happens, the thermal energy of the room's air will then be sufficient to ionize those air molecules and atoms and cause an atmospheric plasma to form inside the room in the approximate shape of the past causative human's circulatory system. As the image forms, the ionization of the air in it will experience a noticeable temperature drop.

At this point, because a local magnetic field in the room has stimulated the formation of atmospheric plasma inside the region of space occupied by the blurry three-dimensional holographic image of the past resident's circulatory system, it becomes possible for this image to become visible.

As the resulting atmospheric plasma particles slowly drift *outside* of the region of the temporarily stable, but still invisible three-dimensional image formed in the room's interior due to ambient air currents, the ions in the plasma will begin to deionize as electrons return in increasing numbers to resume their atomic orbits within the positively charged atmospheric gas ions from which they were previously liberated when they were *inside* of the three-dimensional image. When these electrons "drop" back down into stable orbits within the positively charged air atoms and molecules, they will lose energy which will be emitted as photons of light and a visible glow will be produced. This glow is always of low intensity and best viewed in a pitch dark room, but is also viewable in a room with subdued lighting. It is usually reported has having a pale bluish color.

As for the local magnetic field required in order to make the convergence holographic "ghost" image glow, this can usually be supplied by ferrous building materials that are present in the dwelling although not seen. For example, iron beams in ceilings and iron rods used to reinforce masonry. There may even be large deposits of iron ore or minerals such as magnetite near or under the dwelling that penetrate each room of the dwelling with a magnetic field of sufficient intensity. Sensitive magnetometers can detect these environmental magnetic fields.

It is important to again emphasize that these Category 2 ghosts are always brief in their manifestations because their formations are reliant upon a precise set of *simultaneously* present environmental factors which may only remain at their optimal values for a minute or less.

So, with the above somewhat complicated mechanism, we have the beginning of a rationalization for the manifestation of the stereotypical ghosts occasionally described by witnesses in older dwellings. They are merely due to a past resident of or visitor to the thereafter considered haunted dwelling who experienced a sudden state of intense physiological stress that resulted in his two types of omnidirectional emitted bodily anti-mass field radiations (the steady and pulsating types) altering the microscopic crystalline structures of the building materials in the floor, walls, and ceiling of the room he occupied at the time of his trauma. The ghostly manifestations that result from this process are, therefore, seen to be a natural form of holography. It's important to realize that these Category 2 effects, though frightening, do not represent the souls of the dead who have returned to the dwelling for some purpose. There are variations of this model for ghostly apparitions that can also rationalize the very rare cases in which the visible image appears to move, float in the air, or penetrate walls, but, once again, their treatment is beyond the scope of this general chapter on the subject and the interested reader should see my previously cited work for further details on such cases. Another important feature of the Category 2 ghosts is that they never intelligently interact with a human witness. They appear and disappear and behave exactly as one would expect an image of a slide or of a motion picture film projected on a wall screen to act.

Before moving on to the Category 3 types effects found in haunted houses, I want to briefly deal with the experiences that so-called "psychics" and "sensitives" have when they enter a dwelling some of whose interior surfaces have one of the above described invisible holographic patterns stored in its structural materials.

As these investigators pass through those particular places in such a dwelling, they usually almost immediately become aware of a sensation of "pressure" that they feel on their skin. In these cases, their skins are most likely reacting to the varying intensities of steady anti-mass field radiation that emanate from the holographic patterns impressed into the building materials that surround them. In most cases, these fixed patterns may not have been impressed sufficiently enough in the past or be currently activated strongly enough in order to project a strong three-dimensional invisible image into a room that can then result in the manifestation of the classical luminous ghost by the mechanism as was described above. But, even a poorly impressed pattern that is weakly activated can still produce vague regions in rooms that can affect a true sensitive.

In these cases, the weak holographic images produced in particular rooms tend to be present for far longer periods of time than do the images that can result in the production of frighteningly glowing ghosts. These very blurry and weak images are not capable of becoming visible on their own, but they can still produce rather bizarre symptoms in the bodies and minds of any true sensitive who interacts with them.

In particular rooms, a sensitive investigator may become aware that some event of a traumatic nature occurred there in the past, but he may not be able to describe exactly what the event was. Should the sensitive's body physically enter that region of the room where a weakly formed image is located, he may begin to feel unusually strong pressure on his skin. This effect is easily rationalized by assuming that the incoming steady neutral anti-mass field radiation that then penetrates his body is combining with any natural internal bodily magnetic fields he produces so as to ionize and thereby stimulate various nerves throughout his body. Thus, he may feel pressure on his skin or, if the ionization being induced seriously interferes with nerve conduction deeper within his body, he may have difficulty breathing, be momentarily paralyzed, or even experience visual disturbances.

Sometimes sensitives as well as even ordinary paranormal investigators can locate a region of space inside of a room which, should they insert their hands into the air of this region, will cause them to suddenly feel a noticeable drop in air temperature. Such a region, known as a "cold spot", is most likely due to the room's surface building materials producing and then projecting some small amount of steady neutral anti-mass field radiation back into the room which then penetrates this region of space and combines there with any natural bodily magnetic fields in the investigator's hand. As this leads to the production of the special type of B rich anti-mass field radiation that behaves like the E rich type, the result will be ionization taking place inside of his hand that then stimulates the nerves which control the diameter of small capillaries in its skin. As sudden vasoconstriction takes place in these capillaries, he can get the immediate sensation that his hand is very cold.

A standard mercury thermometer inserted into the region will, however, only show a normal air temperature for the room. In this case the ionization and cooling of the mercury does not take place because there either is no local environmental magnetic field penetrating the region (that is, a magnetic field produced by the dwelling's structural materials and not by an investigator's hand) or, if any local magnetic field is also penetrating the mercury in the thermometer, then it is not strong enough to induce ionization by combining with any weak steady neutral anti-mass field radiation emanating from the room's interior surfaces which pass through the region.

However, on occasion, a thermometer inserted into a region of air inside of a room may show that a temperature drop of several degrees does exist *only* in that region. In these cases, we can assume that a fairly strong, though still very blurry, invisible three-dimensional image is being formed in that region of air and there is *also* a strong local environmental magnetic field present.

Under these conditions, air in the region will be ionized and the energy to accomplish the ejection of electrons from their respective atmospheric molecules and atoms to form plasma will be derived from the thermal energy of the air itself. This will slightly chill the air in that region of the room and, as the air slowly drifts out of the region due to ambient air currents, it will again warm up as its electrons rejoin the positively charged ions they momentarily left. When such "cold spots" are encountered, their approximate location within a room should be carefully marked.

If the three-dimensional image in which the cold spot exists is not too diffuse or blurry, it may, when the room is pitch dark, just barely emit enough light photons as ionized air drifts out of the affected region to be able to be photographed. By using very long time exposure still photography, it may even be possible to produce photographs of the blurry image that are just barely recognizable as those of a human being. That is, one might at least be able to discern where the past resident's head, trunk, and limbs were located when the microscopic crystal pattern caused by the his or her bodily emission of the two types of neutral anti-mass field radiation was initially weakly impressed into the room's interior surface building materials in the form of a unique holographic pattern of reoriented microscopic crystal structures.

Finally, as far as sensitives are concerned, I want to briefly describe what can happen should such a person actually places his entire body into a region of space in a room which contains a well-formed and fairly strong invisible image of a past resident's circulatory system.

In this situation, if the sensitive has his own strong internal bodily magnetic fields and he is positioned properly so as to accurately superimpose himself with the volume of space that was occupied by the past resident, then the investigator may be able to actually *physically* experience the same trauma that the past resident experienced when he originally caused the altered microscopic crystal pattern to be fixed into the room's building materials! Thus, the sensitive might feel a sudden pain in the part of his body which corresponds to that part of the past resident's body that was previously injured. If a past resident experienced a heart attack, then the sensitive might feel crushing pains in his chest. If the past resident was the victim of some sort of violent assault, then the sensitive might feel acute pain in a part of his body which matches that body part of the past resident that was injured.

This bizarre effect is not limited to just physical sensations, however, but can also involve psychological states. In these cases the steady neutral anti-mass field radiation emanating from the room's interior surface structures that penetrate the sensitive's brain can combine with any internal bodily magnetic fields also present there so as to induce ionization of cerebral neurons that will then result in the stimulation of various structures in his brain. Thus, in these very rare Category 2 cases, the sensitive may suddenly feel rage, fear, love, or grief. From the sensations that the sensitive experiences in the haunted dwelling, he may be able to accurately describe events that occurred there hundreds of years prior.

It is important for the reader to realize, once again, that the experiences of the *genuine* psychic or sensitive do not mean that he is in some sort of contact with the spirits of the dead. Rather, these investigators are simply detecting information about past residents of dwellings that has been preserved in a most remarkable manner. In essence, these investigators are simply detecting and reacting to generally weak three-dimensional images formed in the air at a particular location inside of the dwelling. The creation of these images may be rather esoteric in nature and involve a new form of non-electromagnetic particle radiation known as steady neutral anti-mass field radiation, but they are still only impressions or "residuals" left behind by the dwelling's residents or visitors in the past and nothing more.

Next, we must deal with some of the paranormal phenomena of Category 3 that can occur in haunted dwelling. These involve the very rarely encountered cases wherein the effects that are manifesting can be attributed to the presence of *non*human beings within the dwelling. These types of cases tend to repeat the least often, but tend to be more bizarre and mysterious in nature when they do occur. They truly can make one believe in the existence of restless souls who have somehow been trapped between our world and the next.

As most students of ufology will know, there have now been many cases of alleged abductions of human beings by the extraterrestrial humanoids that operate UFOs. While I am comfortable with the idea that *some* of these cases are indeed genuine, I am of the opinion that, perhaps, well over 99% are not "the real thing". Of the very few that are genuine, the vast majority of these occurred when the abductee was out of doors while either driving or camping out under conditions where he was readily accessible to abduction by a landed or even airborne craft. Perhaps, only very rarely, will ufonauts actually enter a home and abduct a person. This is certainly not beyond their technical capabilities though, but such intrusions can pose great risk for alien beings that can not control the environment that they enter. They could possibly be attacked by aggressive pets, counter attacked by armed humans, or even somehow prevented from leaving the dwelling while the police were being summoned.

In general, then, our visitors prefer to perform abduction out of doors in most cases. However, on rare occasions, they do enter our dwellings. If we imagine that the ufonauts who visit Earth are on scientific missions to obtain detailed information about human societies, then it would only seem logical to assume that this would require them to occasionally enter a dwelling, even an inhabited one, to gain this information. While there, they would, as time allowed, probably make an extensive photographic record of the interior, take samples of various building materials that we use, and even take small manufactured items from the house which, upon later analysis, would give them a fairly good idea of our level of technological progress. And, of course, an occasional human being would be somehow rendered cooperative and then either examined in the dwelling or temporarily removed to their craft for a closer examination. In practically all such cases, abductees are returned to the dwelling and any memories of their experience are removed either via a drug or through a form of strong post hypnotic suggestion.

However, I am not convinced that the main purpose of most house penetrations by ufonauts is for abduction, but rather for strict scientific observation of what, to them, is an alien culture. A secondary reason, quite possibly, is to have very limited interaction with any human occupants of the dwelling in order to see how humans react to the occurrence of unexpected events.

With the aid of portable equipment they carry on their bodies in order to achieve various paranormal effects, ufonauts, like the subterrestrial beings discussed in the previous chapter, can levitate themselves, turn invisible, and even momentarily interpenetrate solid structures like floors, walls, or doors. This equipment enables ufonauts to effectively technologically duplicate many of the biological produced paranormal effects and thereby helps our extraterrestrial visitors hide their involvement during dwelling explorations. If this did not happen, then humans might become aware that what was happening was part of an experiment and that would then alter their reactions and result in behavioral observations by the extraterrestrial trespassers that were less scientifically valid.

If a human resident of such a "visited" dwelling awakens in the morning and finds evidence that furniture or other items were mysteriously disturbed during the night, then he might readily conclude that he has a ghost that is haunting his property. If the human occupant happens to actually see items being moved about though the ufonauts responsible are themselves completely invisible in the optical region of the electromagnetic spectrum, then he will most likely be very terrified and convinced beyond all contrary argumentation that his home is haunted by mischievous spirits!

While extraterrestrial intruders do account for a small percentage of the paranormal phenomena that occur in haunted dwelling which are caused by a nonhuman presence, I believe that a far larger percentage are caused by the intrusions of subterrestrial beings.

As mentioned in the previous chapter, there seems to be rather compelling evidence that the upper geological crust of our Earth is inhabited by creatures who are now the remote descendants of the earliest primitive humans that chose to sit out the various Ice Ages by descending into the warmer interior of our planet rather than to risk extinction on its frozen surface. We saw that these beings were actually just living primates who, in their subterranean caverns, were forced to undergo accelerated radiation induced evolution so that they could survive in their new environment. Most of them have probably undergone so much evolution that the differences between their genomes and those of present day surface humanity are now sufficient enough for these beings to be considered almost fully nonhuman. However, there may be a small percentage of those that relatively recently descended (like, perhaps, within the last few tens of thousands of years) into the upper levels of our planet's crust and which are still quite close, genetically, to present surface humans. Most of these creatures are benign and do not wish to interact with surface humanity which they perceive as dangerous.

Unlike extraterrestrials that usually must rely upon portable equipment to generate such effects as levitation, invisibility, or interpenetration, our planet's native subterrestrials have, over the course of hundreds of thousands of years, evolved the natural ability to generate these effects voluntarily. Since the subterrestrial beings of the Earth's crust can not use fire due to the limited supply of oxygen available to them underground, they never developed the kind of technology that they would need in order to defend themselves against the various weapons that surface humans have at their disposal. These creatures must depend upon their evolved paranormal abilities and simple evasion to insure their survival.

However, although they fear surface humanity and generally avoid any contact with us, they are still very curious about human life on the surface and, perhaps, are envious of the various products of technology that make surface life comfortable for humans. Occasionally, I believe, some of these creatures do come to the surface to explore it and enjoy, for a brief time, its beauty. On rare occasion, they will enter a dwelling in search of items manufactured by surface humans that they can take back to their underground realms. Such things as clothing, utensils, and cups seem to be of particular interest to them as well as small items of jewelry and hand mirrors. When such items mysteriously disappear from a dwelling and no other cause can be found, then I would not feel too uncomfortable advancing the "subterrestrial hypothesis" in order to account for the missing items.

Of course, the best place for them to "hunt" for new items is in inhabited dwellings and this increases the risk that they will interact with its human residents. To minimize this possibility, these creatures almost always enter the homes and maintain themselves in a state of optical invisibility. They can easily enter homes by interpenetrating any solid rock and soil that intervenes between the dwelling's basement and nearby subterranean passageways that run under or near the building. They can also simply emerge from the Earth's surface and walk up to a dwelling at which point they can then easily pass through any locked doors or walls. The only type of home security system that is capable of detecting their presence is one which can detect noises or the infrared radiation which is emitted due to their body heat even when they are invisible. Thus, when such systems seem to go off at night for no apparent reason, it may be due to the penetration of a dwelling by an invisible subterrestrial.

As in the case of intruding ufonauts, a human resident may, on rare occasion, see furniture mysteriously being moved about by these creatures in their search for items to steal. Although the subterrestrial may be invisible, the effects of his actions can be quite dramatic. Books may fly off of shelves, doors may slowly open and close, and drawers may slide in and out of bureaus apparently by themselves! On occasion, these creatures may attempt to touch a human being. They may pull on woman's hair or poke people to watch their reactions. Other annoying pranks can include striking sleeping individuals or suddenly yanking sheets or blankets off of them!

Because the preferred mode of communication in the subterrestrial world is telepathy, these creatures may attempt to communicate telepathically with the human residents of a dwelling. Thus, humans may have sudden urges to do usual things or go to various parts of a dwelling for no apparent reason. A sleeping person may suddenly experience a vivid "dream" which is actually a series of images that are telepathically transmitted into his mind by one of these creatures. Only on the very rarest of occasions will a human actually be able to see such a creature when it is visible. The presence of subterrestrials in dwellings can also account for the sudden presence of intense and unpleasant odors. These smells seem to be ones that would be associated with a creature that inhabits the depths of the Earth's crust and often have an oily or sulphurous quality to them.

Like ufonauts, the subterrestrials who enter human dwellings tend to be benign, although their presence is usually unwanted and, when initially experienced, can be terrifying to human residents. Knowing the nature of this category of effects that are experienced inside of human surface dwellings does not necessary remove the fear humans will experience. Many owners of such dwellings may find the visitations of these subterrestrial creatures so annoying that they will simply sell the building and move. This is often the case if their visitations are perceived as dangerous to the health or safety of the human residents. Others, convinced that the creatures are the earthbound souls of deceased relatives or former residents, may simply tolerate the phenomenon and try using various religious rituals in an effort to help these supposed souls "pass over" to a "better place". In many cases, the visitations will just suddenly stop. When this happens, I suspect that the particular subterrestrial involved just got bored with his regular visitations or has ceased to find any additional items in the dwelling which he might purloin. Perhaps others in his subterranean world learn of his boredom and they then also avoid the dwelling which can give its human occupants a welcomed period of peace.

Category 4 includes those cases in which people report paranormal events inside dwellings that never actually occurred or create evidence that they then report as having a paranormal origin. I debated with myself whether to include such hoaxes as a category in this chapter because they are more a product of human psychology than reality, but decided to do so since they occasionally occur and the serious researcher should be aware of them lest he waste his limited resources trying to document such alleged events.

Understanding the exact motivations of a person who would *deliberately* claim a nonexistent paranormal experience and then present bogus evidence to back up such a claim is a somewhat difficult thing to do. (I say "deliberately" here in order to distinguish these individuals from those who may have honestly misinterpreted certain events and experiences which they believed to be real or who, due to a serious mental health condition, may have hallucinated them.) Consequently, I can only present my opinions here about the matter because there has, to date, been little serious research concerning it.

Basically, I believe the primary motivations for committing such a hoax are fourfold:

a.) the hoaxer may actually have had a real paranormal experience in the distant past that was never reported at that time for one reason or another. He believes that it will do no harm if he merely "updates" the experience at the present time, along with "evidence" to support it, so that it can finally be documented and that this may then actually serve some sort of beneficial effect; that is, it will increase interest in the subject and thereby stimulate serious research into it that might benefit humanity. He then uses the perceived admirable intention of his act to relieve himself of any guilt for having committed it.

b.) the hoaxer, perhaps being bored, is looking for some attention which he will certainly get when paranormal investigators arrive to interview him, view his fabricated evidence, and perhaps even place his residence under a nightlong surveillance. He then gets gratification by being the center of attention and designated as one who has been fortunate enough to make contact with a "higher" level of reality. This is somewhat similar to the "Munchausen's Syndrome" that is seen in the medical world in which people falsely claim various symptoms in order to gain the time and attention of trained medical personnel.

c.) the hoaxer may have some sort of deeply held philosophical or religious beliefs that he is expressing and seeking to reinforce by committing the hoax. Thus, he may believe in the existence of a spirit realm or an afterlife and the hoax allows him to advertise this and, perhaps, get others to adopt the same believe system as he. He may believe that spreading his belief system will somehow lead others to "salvation" by "saving" their souls and this relieves him of any guilt for perpetrating the hoax.

d.) the hoaxer may obtain psychological gratification by committing hoaxes and fooling educated people. His actions could be his way of compensating for his own feelings of inferiority due to a lack of education or social acceptance. After all, if he can fool dedicated, serious, and intelligent people such as paranormal investigators with his hoax, then that, to him, means he is smarter than they! A similar motivation is seen by those known as "hackers" who spend countless hours at their keyboards trying to find vulnerabilities in computer software programs or systems that they can then exploit to cause havoc in the lives of others or make a financial gain for themselves.

With this information, the paranormal investigator should be better prepared to sort reality from fiction as he attempts to document and analyze the events and experiences occurring in haunted dwellings. Today there are a variety of electronic devices available that can be used to analyze the spoken testimony of witnesses and determine, with a fair degree of accuracy, whether or not they are being truthful. Such means should be utilized whenever possible. Since, rarely, will a hoaxer ever admit to their hoax, it is generally not a good idea for investigators to confront him if a hoax is suspected or even if it can be proven without a doubt. In such a case, the investigators should make some excuse and then withdraw from the scene as soon as possible. Many hoaxers, if not getting the feedback they anticipated from one team of investigators, will then simply try to gain the attention of other investigators. In time, if they experience continuing frustration, they may finally stop their behavior and turn to other interests.

As this chapter concludes, I want to briefly touch upon a final cause of dwelling hauntings that may be due to the presence of a *human* being who was previously deceased, but who has, somewhat miraculously, actually been restored to life! This possibility is so bizarre that I am even hesitant to include it here, but the evidence for these extremely rare Category 5 cases does exist and they are undoubtedly the most interesting possible cause for a haunting.

By treating the matter of a human who was previously deceased, I am not referring to the many reported cases of "zombies" that surfaced during the twentieth century in various Caribbean islands where voodoo is practiced. These cases only *appeared* to involve the return of the dead. In reality, they involved unfortunate people who had been poisoned with an extract derived from the puffer fish which produces a temporary dead-like state of coma.

After the victims were promptly buried on the same day they died in the warm tropical climate, they were then quickly unearthed by the individuals responsible for their poisoning and revived to consciousness. However, the deep coma that they were in for several hours coupled with the greatly reduced supply of oxygen to their brains would result in a loss of almost all memory of their personal identity and past life (note that a victim must be exhumed within twelve hours of burial or he will die from hypoxia in a sealed casket). In this brain damaged state, they would then be sold and used as slaves to mindlessly perform work for their masters. Only a handful of the victims of this process have been successfully restored to their prior states. Again, it is important to realize that these zombies were not resurrected to life because, in fact, they were not really dead at any time.

The Category 5 cases I am referring to, however, involve a person, usually a young adult, who suddenly appears in the locale of his burial a few weeks or months *after* he has *actually* died! In most cases such individuals seem to then possess enhanced paranormal powers that allow them to display the same types of effects noted for some of the extraterrestrial and subterrestrial beings previously described that involve the ability to levitate, turn invisible, and interpenetrate solid barriers. These resurrected humans may be seen at night on the roadside signaling to lone drivers for a ride and, if given one, then will mysteriously disappear into thin air at the conclusion of the ride. They may visit their former residence or its neighborhood and attempt some sort of conversation with associates there that they knew in their former life. Needless to say, anyone who might encounter such a person that everyone knows to be deceased would surely be 100% convinced that he had seen a ghost!

There are alternative explanations for such bizarre Category 5 cases. Obviously, one can just dismiss the anecdotal "data" of these cases as being false; that is, that those reporting them were mistaken about the identities of the persons spotted or just lied about the incidents having taken place at all. Or, one can advance another hypothesis that acknowledges the reality of the reports, yet does not require one to believe that they are due to the materialization of a formerly living person's "restless" soul. Let me now briefly give an outline for this last alternative approach to rationalizing such cases.

I begin by stating that the source of these cases was a person who *did* actually die and was buried. Let us further assume that she was a young adult female and died of some infectious disease that overwhelmed her body's natural defenses until she finally expired and, thus, there was no gross physical damage to her bodily tissues at the time of death. Let us further assume that her cadaver was *not* cremated, but, rather, either was then drained of blood as it was embalmed with formaldehyde fluid or, as was more common prior to about the beginning of the 20th century, simply given a quick burial before significant putrefaction began to set in which is, depending upon climate, within about three days if refrigeration is not available. The latter method of burial might still, of course, take place even today in most countries if that is the desire of the deceased's survivors based upon their particular religious beliefs or family burial practice.

Now let us imagine that the graveyard in which this newly deceased person was buried was near a subterranean passageway that subterrestrial beings use as they move between their underground world and the surface of our planet. Since the passageway runs under the graveyard, it is possible for these creatures to physically interpenetrate any intervening rock and soil between their passageway and a newly buried casket. They might do this occasionally to see it they could recover personal items from the casket that had been buried with the deceased. These items would be things such as clothing or jewelry.

In the particular generalized scenario being considered here, we can imagine that, as a subterrestrial being reaches the female's casket (by the "swimming" method described in the previous chapter), he manages to touch it and then willing cause *various* excess anti-mass field radiation emissions from his body's circulatory system (which are a mixture of steady neutral and B rich anti-mass field radiations) to greatly lower the mass of the casket while simultaneously greatly lowering the *repulsive* electrostatic forces that exist between the *like* electrically charged subatomic particles in its various structural molecules and between those subatomic particles and the like electrically charged subatomic particles of the molecules in objects *external* to the casket.

The reduction in these repulsive forces, however, is not total and it is still possible to use the repulsive electrostatic forces existing between the subatomic particles of the casket and objects external to the casket to then apply some small amount of driving force to the casket. This condition of the casket of greatly increased, but *not* complete interpenetrability allows the subterrestrial to grasp one of its handles, slowly tug on it, and then literally slowly pull the entire casket through the intervening feet of soil and rock until it reaches the air filled subterranean passageway under the cemetery.

At this point, the subterrestrial releases the casket and, as it floats away from him, his bodily anti-mass field radiations no longer affect it so that it begins to slowly regain its normal properties of mass, weight, and lack of interpenetrability. As it does so, it slowly sinks to the floor of the passageway.

While resting on the floor of the underground passageway with its full mass again, the casket's cover is forced off and the subterrestrial then views the body of a young woman. Let us assume that the subterrestrial is one from a race that inhabits the upper levels of the Earth's crust and is still fairly human in appearance due to a genome that is not yet too much mutated from that of surface humanity. The untimely dead of the attractive young woman seems tragic to the subterrestrial and, rather than plundering her casket, he decides to initiate a very rare procedure: the actual resurrection of a deceased human being by a subterrestrial!

It is, unfortunately, beyond the scope of the present chapter to go into all of the many physical processes that could make the resurrection of the dead possible (the reader interested in more details on this fascinating possibility should see the chapter titled "Alien Medical Technology and Immortality" in the author's book *The How and Why of UFOs*), however, let me just state that several subterrestrial beings, working together, could provide the necessary conditions for resurrection to occur. Several of the creatures would have to surround the body of the deceased and then each would produce their maximum output of steady neutral anti-mass field radiation. This radiation would overlap in the region of space occupied by the female's cadaver and would immediately lower its mass to zero. Since the kinetic energies of mass possessing subatomic particles *must* be conserved as their gravitational and inertial mass dwindles in magnitude, they will respond by accelerating toward infinite velocity. Indeed, impossible as it might seem, their velocities will actually become infinite when they do finally achieve total masslessness! (In such a state a particles possesses no mass as far as gravity and inertia are concerned, but still possess all of the energy associated with its original unnegated mass.)

As the velocities of all of their formerly mass possessing subatomic particles (mainly the electrons, neutrons, and protons) become infinite, all atomic and molecular motions within the dead woman's body will also become infinite. This will then result in a dramatic increase in the rate at which various biological and physical processes in her cadaver would normally take place.

For example, her cadaver, while completely massless as far as gravity and inertia are concerned, will rapidly absorb water from the dank atmosphere of the subterranean passageway in order to replace any lost from it since the time that she died. With newly rehydrated body fluids containing the correct concentrations of electrolytes, her heart will spontaneously start to beat again and all of her body's natural healing processes will be reactivated and enormously accelerated. Any infection in her system will be destroyed by her reactivated immune system and any cellular damage caused by the infection will be quickly replaced with fresh living cells. If the body had been embalmed, the enhanced action of the liver would rapidly break down the circulating formaldehyde fluid and convert it into harmless compounds. Any blood the embalming fluid had displaced would be quickly replaced with healthy blood cells by her bone marrow. In, perhaps, less than a minute after the subterrestrials begin the joint and overlapping emission of their maximum output of steady neutral anti-mass field radiation, the young woman will begin breathing and then finally regain full consciousness!

Initially, she may believe that she is still at home in her sick bed and be wondering why it is so dark. As fear begins to set in, the subterrestrials will then use their powerful telepathic abilities to tell her to be unafraid and will put her into a tranquil mood. However, she will still be in the dark and wondering where she is.

I am of the opinion that the "resurrection process" may have an interesting secondary effect. If performed on a human, it may bestow limited paranormal abilities onto that person. If this is the case, then the hypothetical human female we are considering here may find that she now has vision that extends down into the near infrared region of the electromagnetic spectrum and that she is finding it easier and easier to pick up the thoughts of the benevolent figures that surround her. She can see that she is in some sort of cave and that the beings near her seem human. But, she wants to know exactly what is going on and how she got to her new location from her sickroom.

Perhaps over the next hour or so, the subterrestrials would telepathically communicate to her that she was dead and that they brought her back to life. They would explain to her all about their subterrestrial race's existence and their lives underground. They would also tell her that, because of this and her new powers, she should not try to return to her former life on the surface of the Earth. Her sudden appearance after being buried would just create too much shock in the family and community to which she might want to return. Although her friends and family would be overjoyed to see her alive again, the news of this event might leak out and other humans, desperate to obtain a means of escaping death, would have the entire cemetery excavated until they found the passageway used by the subterrestrials. If these humans then discovered the subterranean dwelling places of the defenseless subterrestrial beings who had resurrected this woman, they might threaten or even kill the subterrestrials in order to be able to overcome death!

After thinking the matter over calmly for several days, the young woman might finally realize that she could not really permanently rejoin surface humanity, but would have to settle for an occasional visit to the surface location of her former life. She might visit her former home and perhaps even enter it to retrieve precious items to take back with her to her new subterranean life. When these items were noticed to be missing by her friends or relatives, they would assume that it was a "sign" that the deceased young woman had taken them with her into the afterlife. In a sense, they would be right, but it would not be an afterlife up in heaven. It would be an actual *physical* afterlife deep within the protective crust of our planet.

There is even a more compelling reason why the hypothetical woman in this generalized scenario would not want to return to the surface world to live. With enhanced paranormal powers, she would now have the ability to voluntarily produce anti-mass field radiation from her own circulatory system just as the subterrestrials do. One of the consequences of the massless state is that, aside for allowing for the possibility of the resurrection of the dead, it also allows a living being to indefinitely rejuvenate its bodily cells and, thereby, attain actual physical immortality! If the woman was to return to the surface world and try to live a normal life there again by not revealing her experience and abilities, she would find that all of her closest friends would slowly grow old and die off while she would forever remain young and vital. Thus, as the centuries passed, she would only know the continual heartache of losing the people closest to her. Unfortunately, this is the price demanded by all those who want to escape the finality of death.

But the fate of our hypothetical resurrected human female is not that bleak. Life could probably be comfortable for her in her new subterranean environment. She would have the company of creatures with the same powers as she. There would be food, drink, and social interaction. She might even fall in love with a male of the more human looking subterrestrials and bear children. There would always be visits to the surface world and the possibility that the day might come when it would be possible for her to openly return to the surface again.

Such a return could take place shortly after the day when humanity discovers how to artificially generate anti-mass field radiation of its own and begins to realize all of the miraculous effects that can be achieved with it. Soon thereafter, surface humanity would have the same abilities that the subterrestrials now naturally possess. When that happens, there would be no reason for surface humans to try to harm the subterrestrials and, in fact, the subterrestrials would be able to openly show themselves. I am confident that the surface humanity of that day will not react with fear, but, rather, with fascination about the existence of humanity's long lost subterrestrial relatives.

I can envision a day when surface humans, subterrestrials, and even extraterrestrial races will be able to peacefully and beneficially interact with each other. Perhaps that day will dawn sometime in the next few centuries.

Chapter 5

Decoding the Great Seal of the United States

I HAVE ALWAYS HAD A FASCINATION with the symbols used in heraldry which include the various ones found on coats of arms, seals, coins, stamps, flags, and even national monuments. The symbols used are all intended to convey the heritage, ideals, and / or aspirations of whomever initially commissioned the design. The meanings of the symbols may be overtly spelled out in the design or the symbols may be ones that should be recognizable to the audience for whom the symbols are intended. On rarer occasion, however, the symbols may contain hidden information which has been carefully encrypted within them. This information may not be intended for a contemporary general audience of the symbols, but may actually be intended for a special future audience. The encrypted information may be a prophecy about future events or even a warning to future generations who will be able to properly interpret the symbols.

In this chapter, I want to delve into certain unusual numerical information I noticed, several decades ago, while studying the Great Seal of the United States. This is the set of symbols found on the back of the basic unit of United States currency: the ubiquitous one dollar bill. I am now releasing this information because, although I have seen much information and speculation about the symbols in the Great Seal, I have not seen anybody else provide the specific numerological analysis that will shortly be presented. First, however, for the general reader unfamiliar with this subject, a little background information is in order.

On July 4th of the year 1776, the Constitutional Convention of the newly formed United States of America (which only consisted of its former original 13 British colonies at the time) convened and 12 of the

13 colonial delegates present approved the wording of a "Declaration of Independence" to formally announce their intention to have a separate political and economic existence from Great Britain.

There continues to be some controversy as to exactly when the United States of America became a separate political entity from Great Britain. For example, it was on July 2nd, 1776 that the "United Colonies of America" officially changed its name to the "United States of America" when the delegates of the United Colonies of America adopted a resolution by Richard Henry Lee and John Adams to sever ties with Britain. However, it was on July 4th, 1776 that the official wording and date of this declaration was approved by the majority of the delegates. This dated document had been, by the evening of July 4th, 1776, only signed by the President of the Congress, John Hancock, and the Secretary of the Congress, Charles Thomson, but copies of this document, dated July 4th, 1776, were ordered printed for distribution to the newly declared states. It was not, however, until August 2nd, 1776 that the July 4th, 1776 declaration was finally signed by all members of the Congress. I, like most Americans, will accept July 4th, 1776 as the "official" date that the United States came into being as a separate nation from Britain. I do this because that date appears on the copies of the document used to inform the citizens of the new nation about the event and it was a copy of this *same* partially signed document which later in the year 1776 was sent to King George III to inform him of the newly self-declared status of Britain's former colonies.

On that historic day of the 4th of July, 1776, Benjamin Franklin, John Adams, and Thomas Jefferson were given the job of designing the Great Seal of the United States. In years past, I recall reading somewhere that all three of these men were practicing Freemasons and as members of this international society, incorporated secret ancient symbolism into the design. I also recall reading somewhere years ago that this suggested that the political system and destiny of the United States was somehow being controlled by secret human societies whose ultimate plans for the people of the United States could only be dimly perceived by a carefully study of the Great Seal.

Ordinarily, I put little faith in such conspiratorial speculations. In all cases I have researched, the evidence for them is a matter of subjective interpretation with no actual proof ever becoming available. The future predictions made from such speculations inevitably fail to manifest, but in those many cases, the original prognosticators have usually long since died off by the time that the original prophecies were scheduled to be fulfilled. If they are still alive and questioned about the failure of their predictions, they will simply state that their original speculation was faulty due to some previous misinterpretation of events or symbols and that it must now be reformulated. More commonly, they may assert that their predictions are correct, but require more time to be fulfilled. Eventually, a new generation arises and may continue the myths and predictions of the former one. But, nothing really changes over time. Certain conspiracy theories eventually become mythologized and develop a life of their own. Therefore, in this chapter, I will not waste space tracing all of the various possible cultural symbolisms found in the Great Seal of the United States, but rather will focus on the very few unusual aspects of it that do seem to me to have some relevance for the times in which we live.

In order to finally get the Freemasons out of my treatment, let it suffice to say that, at the time that the Great Seal was commissioned, only Benjamin Franklin was a practicing Freemason and there seems to be recent evidence that neither Adams nor Jefferson were Freemasons. At this time I do not feel that the Masonic connection to the Great Seal's initial specifications is strong enough to justify treatment herein. However, this does not necessarily mean that I do not believe that some covert group or individual with intimate knowledge of certain events that occurred during the Revolutionary War did not influence the later versions of the seal, particularly its reverse side. With that matter out of the way, let us proceed with the analysis.

The *original* design for our Great Seal was not finalized and approved for almost 6 years *after* its commission in 1776. That final approval occurred on June 20, 1782. It was a slow process hampered by delays and even the "final" design was somewhat general and went through many variations over the next 150 years! All that was really finalized on that day in 1782 was the blazon for the great seal. This blazon is only a technical *description* of the seal made in language that is in accord with the rules of heraldry. It is not really a final design and, in fact, allows for great variation in the pictorial representation over time so long as the basic elements of the blazon's description are present. Most past writers on the subject did not seem to me to be aware of the fact that the presently used Great Seal which appears on our dollar bill was not engraved until 1935! Thus, most (but not all) of the design elements found in our current Great Seal's design cannot be used to draw conclusions about the events surrounding the earlier designs.

For example, much is made of the observation that the number 13 shows up several times on both the Seal's front or "obverse" side and its back or "reverse" side (note that *both* sides of the seal appear on the right and left sides of back of the US dollar bill). While it is true that the Hebrew word for the number 13 is the same word used for God in the Old Testament (which is *"Adonai"*), I believe this is just a coincidence and the number is really meant to refer only to the original thirteen American colonies whose ideals and aspirations are expressed in the Seal's design. The design was originally intended to convey to the world that the United States was a new political entity that would endure by following Judeo-Christian moral principles. The excessive usage of groups of 13 objects in our current design is not seen in the Seal's very earliest renditions. They seem to have "crept" into the design during the 19th century.

Before continuing with my analysis, let us very briefly summarize the heraldic elements that were originally specified in the Seal's blazon in 1782. For the front or obverse side an American bald eagle is specified. This eagle's beak holds a long, narrow ribbon with the Latin words *"E Pluribus Unum"* (which translates into the English "One from Several"). The Latin phrase contains 13 letters. In the eagle's right (or "dexter") claw is an olive branch and in its left (or "sinister") claw are a group of arrows (note: only in later designs are there 13 olives and arrows shown). These represent the new union using either the power of peace or war to secure itself. A shield with 13 red and white stripes is required which are located immediately below the top blue section of the shield. The stripes on the shield were intended to represent the original 13 states that supported the President and Congress which were represented by the top blue portion of the shield. As far as the colors are concerned, red represents valor, white represents purity, and blue represents justice. The design of the 13 stars above the eagle's head is arranged in a compact geometric pattern that resembles the Star of David. Again, I believe this is a coincidence because this pattern is the most symmetrical one into which 13 stars can be arranged.

We now, finally, come to the back or reverse side of the Great Seal and, once again, the reader must bear in mind that its design actually took about 150 years after 1782 to be finalized and engraved as shown on the back of today's US one dollar bill. I can only assume that it is a composite drawn up from several of the designs made over the years since 1782. It shows a 13 step pyramid with the number 1776 carved into its base course of masonry in Roman numerals. It is a "truncated" pyramid meaning that it has a flat top and its capstone is missing, but in the present day engraving this has been replaced with a hovering and radiant eye known as the "Eye of Providence". The pyramid represents the hoped for endurance of the newly formed United States of America by its citizens and the eye (which is the favorable right or dexter eye) represents God's favor and protection being directed down to the new union. This aspiration is reinforced by the Latin words "*Annuit Coeptis*" above the pyramid which in English means "He favors our undertakings". A long scroll beneath the pyramid contains the Latin words "*Novus Ordo Seclorum*" which translate into the English "New Order of the Ages" and signify that an "American era" has started beginning in the year 1776. However, it is the pyramid itself which is the focus of this short chapter and which must now be treated.

Several decades ago, I decided to use a 10x power jeweler's loupe to take a good close look at the Great Seal's reverse side pyramid engraved on the back of the US 1 dollar bill and see if I could find anything interesting about it. In particular, I wanted to see if I could accurately determine the number of stone blocks from which the physical structure it depicts would be constructed in reality and found this to be a very difficult thing to do. The problem is that, while the horizontal lines of the 13 levels or courses of masonry are engraved rather distinctly, the individual stone blocks within a course are somewhat difficult to make out because the vertical lines representing the grooves between the stones are sometimes missing or are obscured by the vertical line shading that appears on the front side of the pyramid. I think that the first time I worked on this analysis, years ago when my eyesight was still very sharp, I determined that this pyramid contained exactly 300 stones blocks. I believe that I originally discounted the pyramid's base (the lowest level with the Roman numerals engraved into it which, at that time, I thought might be a solid, *one* piece slab of stone) and considered *only* the upper 12 courses of masonry that rested on the base.

I observed that the number of stone blocks varied from course to course, but *averaged* out to about 25 stones per course. I noticed that these stones varied only in height *and not in width or depth* (the only exception to this rule being the smaller blocks used to complete the corners of each course). I assigned the lowest and first course of stones (which rest directly on the one piece base) an arbitrary height of 1 unit and that then gave the uppermost and final 13th course of stones a relative height of only 0.5 units. The "average" stone block would therefore have a relative height of 0.75 units and its volume would, therefore, be 0.75 that of the largest first course stones since their widths and depths were the same.

With this information, one could then quickly determine the total volume of the pyramid as a multiple of the volume of one of these average stone blocks by multiplying 300 blocks by 0.75 cubic units per average block. When this was done, I got a figure of 225 average blocks which only means that the pyramid was 225 times more voluminous than an average sized stone block within it or could be said to contain enough material to make 225 average stone blocks. I then noticed that the *height* of an average block (which was 0.75 of the height of a first course stone or 0.75 units) reminded me somewhat of the *date* that appears on the Declaration of Independence which is 07/04/1776. Because of this, I then decided to use a new figure, 0.74 cubic units per average block (from the declaration's date being the 7th month and 4th day!), for the volume of an average block of stone in the pyramid and, upon multiplying it by 300 stones, got a total of *exactly* 222 average sized blocks contained within the pyramid. I thought that this was no accident and was, in fact, incorporated into the pyramid's dimensions to signify something.

Year later, I wanted to see if there was another mathematical method that could be used to extract this unusual number from the Great Seal's pyramid and decided to try a direct measurement technique on the engraving of the pyramid shown on the back of the US 1 dollar bill. This time, a more careful examination of the pyramid's base showed me that, indeed, it was not a single slab of rock as I had previously believed, but, like the 12 courses of masonry it supports, it, too, is made up of stone blocks. Some of the grooves between the base course stones can be barely made out above the Roman numeral M and between the two C numerals. In this alternative analysis, I would work with the full 13 courses of stone blocks in the Great Seal's pyramid which rose right from the ground up to its flat top.

Since the 7th course of stones is exactly midway between 6 upper and 6 lower courses of stones, it must be considered to be made up exclusively of average sized blocks of stone. Using a clear plastic ruler's millimeter edge and my trusty 10x power jeweler's loupe, I measured the width of the 7th course stones and found it to be about 9 mm wide. The height of the pyramid from the ground up to its flat top was about 11 mm. With these measurements, it is now a simple matter to obtain the volume of the pyramid in cubic millimeters. This volume is equal to its height times the width of the average or 7th course of stones squared. So the total volume of the Seal's engraved pyramid is 11 mm x 9 mm x 9 mm which equals 891 cubic millimeters.

By measuring several of the most clearly defined blocks in the courses *nearest* the 7th course of stones (the *widths* of the stones in the 7th course itself are virtually impossible to discern), I estimated the width of an average block of stone in the 7th course as about 2mm and its height as about 1 mm. Thus, the volume of an average 7th course stone block is about 1 mm x 2 mm x 2 mm which equals 4 cubic millimeters. Now we can determine how many average sized stone blocks are contained in the Great Seal's pyramid by simply dividing the engraved pyramid's previously calculated total volume of 891 cubic millimeters by the volume of one of its average sized blocks of stone which is about 4 cubic millimeters. Doing this gives a figure of 222.75!

Since the dimensions I used in my calculations to obtain the engraved pyramid's volume were not exact (the actual ones are just a tiny, tiny bit less and not accurately measured with a ruler), I believe that the actual figure would be 222 average stone blocks contained in the Great Seal's pyramid. When repeating digits like 222 emerge from the geometry of a heraldic symbol, they often represent a time period, usually years. So, the 222 average sized blocks contained in the structure of the Great Seal's pyramid most likely represent a period of 222 years which began in 1776. This number, however, could also represent a time period of 22.2 years or even 2.22 years.

In trying to determine the possible significance of the number 222 derived from the pyramid of the Great Seal of the United States, I decided to research some of the personalities that were prominent around the year 1776. I needed look no further then the history of George Washington who is permanently linked to the Seal's pyramid by having his portrait displayed on the front side of the US one dollar bill.

Washington was born in 1732 and died in 1799 at the age of 67 after being treated for a bad flu by being bled excessively by doctors that did not know any better. Early in life this robust, 6 foot 4 inch tall, 225 pound, intelligent blue-grey eyed man became a wealthy landowner. Although he was an accomplished surveyor, he was also devoutly religious and would think nothing of stopping his activities and then dropping to his knees to offer a prayer of thanks to God or one for guidance from Him. His tall stature and personal integrity made him a natural leader and it is no wonder that when the Revolutionary War finally got started he was quickly appointed Commander-in-Chief of the "Colonial Forces" by the Continental Congress.

However, the year of 1777 did not go well for General Washington's campaign against the British forces. Money and supplies that had been promised to him by the Continental Congress had not arrived in time and his army was slowly being pushed out of the state of New Jersey (by the way, Washington spent fully half of the Revolutionary War fighting in my home state of New Jersey). By the winter of 1777, he had been forced to retreat to Valley Forge in the neighboring state of Pennsylvania. His winter encampment at Valley Forge was a miserable one, indeed. His men were half starving and lacked adequate clothing to protect them against an unusually frigid winter. Washington, however, managed to keep morale up and even began using his own money to pay for supplies. During this time of despair, he often visited a secluded thicket and prayed for divine aid. He knew that the price of losing his campaign against the British would be the end of the newly formed union of states as well as his own execution! In 1777 there were only about 3 million people in the United States of America and they were bitterly divided over the matter of whether or not they should break away from Britain and form their own nation. At most, only about 3% or 30,000 of the colonists were directly involved in the Revolutionary War.

One afternoon during that bitterly cold winter at Valley Forge, Washington had an experience which he would remember for the rest of his life and that he would later only discuss with his closest associates. He was in his quarters doing paperwork at his desk when suddenly he became aware of the presence of an attractive woman standing in the room with him. He asked her several times who she was and what she wanted, but in answer she only raised her eyebrows and transfixed him with her gaze.

Unable to move or speak, General Washington noticed that the background around her started to blur and change into a large panoramic map of the North American continent, the Atlantic Ocean, and the continents of Europe and Asia. After a few moments, a dark "angel" appeared over the Atlantic Ocean portion of the tableau he viewed and began creating dark clouds that slowly drifted over the newly formed Union. Washington was told by the mysterious woman that this represented the war he was then engaged in with the British. He was also shown images that represented, according to most modern interpretations of the symbols, the Civil War that would divided the nation in the 19th century and the 20th century struggle between the United States and the communism of the former Soviet Union. He was assured that, despite all of these crises that would befall the United States, the Union would survive and continue indefinitely. Then, as quickly as the mysterious woman had arrived, she vanished!

I have only given the barest summary here of what has come to be known as the "Washington Prophecy". The entire incident was described in a letter in Washington's own handwriting that he sent to Jefferson shortly before Washington's death in 1799. This letter in now in the possession of the Library of Congress and, because of its length, I will not include it in this chapter. The reader interested in it can find it reproduced on various internet websites devoted to prophecy.

After Washington's amazing "vision", eye witnesses saw him emerge from his quarters with his face drained of its color. He immediately secluded himself in his private thicket to pray and after an hour or so, returned to his camp and told his generals of his bizarre experience. He told them that his experience had to mean that the Colonial Forces, despite their recent setbacks, were destined to defeat the British. Washington and his generals decided to plan their counter attack against the British to take place as soon as possible and not to wait for warmer weather.

In the early morning hours of December 26th, 1777 Washington's army began crossing the Delaware River that separates the states of Delaware and New Jersey. By 4 a.m. of that freezing cold morning, he had landed along with his 2400 men and their supplies at Mackoney's Ferry which is about 9 miles north of Trenton, New Jersey. By 6 a.m. he was attacking the British forces in Trenton which mostly consisted of Hessian mercenaries (hired by the British to bolster their forces in

New Jersey) who were still groggy from the previous Christmas Day's festivities. The entire battle only lasted about three quarters of an hour! In the process 200 Hessians escaped, over 100 were killed, and 950 were captured. General Washington only lost 2 men in the combat, 2 others froze to death, and a few dozen were wounded. In the following days, the captured Hessian soldiers were paraded through the streets of Philadelphia, Pennsylvania and the citizens there gave thanks to God for the victory.

In short order, Washington went on to win major victories all across New Jersey, but the battle of Trenton was *the* pivotal battle of the Revolutionary War. It was his victory there that swayed the vast majority of colonists in favor of independence from Britain and of support for the war. It is important for the reader to realize that this "make or break" battle probably never would have taken place if Washington had not had his "vision" wherein he was shown that he would not lose the war! Most of the secular biographers of George Washington tend to downplay his vision or dismiss it as a dream that he had after dozing off at his desk. I, however, think this would be unlikely. Dreams do not have the length and detail that he described in his letter to Jefferson about the incident. Also, the idea that he would have invented the experience would have been totally out of character for him. I further dismiss even the possibility that his vision was a hallucination.

Washington was a healthy, active man who never displayed any prior symptoms of mental illness. I am, therefore, forced to conclude that he had a real, physical experience. A being appearing to be a human female did enter his quarters and was able to immobilize him using telepathy. She then delivered a message to him telepathically which his mind received as both words and a series of images. I can only conclude that this being was, because of her advanced powers, not an ordinary human being. The only possibility I can see for her origin is that she must have come from somewhere *outside* of our Earth! I realize that this assertion may seem outlandish to some readers, but it is the only one that fits the facts of the incident. While Washington would have considered her an angel sent from God, I and others might consider her an agent sent by some extraterrestrial force that maintains surveillance on Earth's humanity and takes an active role in shaping the destiny of our planet for some, as yet, unknown purpose. Whatever one's beliefs are, the Washington vision is dramatic proof that some outside agency, whether spiritual or physical, has and most likely still is involved in controlling events on our planet.

At this point we can make a possible connection between the number embedded in the dimensions of the Great Seal's pyramid and the Washington Prophecy. I feel that it is most likely the case that the pyramid's 222 average stones represent a time period, in *years*, during which the events depicted in the vision General Washington received that cold afternoon in 1777 will take place. If so, then his vision covers the time period from July, 1776 to July, 1998. During this time, the United States of America prevailed during the Revolutionary War and its brief revival during the War of 1812, the Civil War, the Spanish-American War, World War I, World War II, the Korean War, the Vietnam War, and a handful of smaller conflicts after those including the first war ("Dessert Storm") against Iraq caused by Sadam Hussein's invasion of his neighboring state of Kuwait and the potential threat this represented to other oil rich Persian Gulf states.

The number 222 derived from the Seal's pyramid could also represent a shorter period of 22.2 years or even a shorter one of 2.22 years. It is interesting to note that Washington only lived about 22.2 years after the approval and partial signing of the Declaration of Independence in July of 1776 until his death in 1799. He was President from 1789 to 1797 and only lived about 2.22 years after leaving office. I also noticed that Washington was born in 1732, a year whose digits add up to 13 and he died at age 67 which, again, is an age whose digits also add up to 13. The year he died was 1799 whose digits add up to 26 which is 2 times 13. Is it any wonder that he would become President of the first 13 states?!

There seems to be some sort of loose affinity between the destiny of the United States and the number 222. If this number represents the time period covered in the Washington Prophecy, then that time period ended at a significant time which was in July of 1998. That time was significant because it was approximately the time that Osama bin Ladin and his Al Quaeda terrorists decided to declare war on the United States. Perhaps their leaders were aware of the Washington Prophecy and also managed to decode the pyramid on the Great Seal! They may then have figured that the promise of protection given in the Washington Prophecy was scheduled to end in July of 1998 and because of that, this time would therefore be the best for them to openly declare their war against the United States. They may have reasoned that a declaration at that particular time would best serve to increase their chance of realizing ultimate victory against the United States which is the chief obstacle to their dream of establishing a worldwide "Caliphate" based on their interpretation of the religion of Islam. Perhaps they missed the part of the Washington prophecy that states that the Union (or United States of America) would endure *forever*.

In reviewing the horrific events of November 11th, 2001, usually referred to as just "911", when the World Trade Center towers were destroyed by Al Quaeda terrorists using hijacked passenger jets, I recalled how the two towers reminded me of the number 11 when viewed from a distance. There were *two* buildings that were destroyed that together looked like the number 11 and 2 times 11 equals 22! It is also interesting to note that the first attack on the World Trade Center actually occurred back in early 1993 when a powerful truck bomb was detonated in the parking levels beneath one of the buildings. If we consider the individual digits in the year 1993, we see that, when added, they also sum up to 22! If we consider the year when the World Trade Center towers were finally destroyed, 2001, we can treat its first two digits (starting from the left) as being the number 20 and then add the last two digits, 01, to it to obtain 21 which is close to the number 22. Something even more interesting happens when we use the month and date of the second successful attack. September 11th was the 254 day of a year with 365 days in it. We can get the number 254 from the number 365 by subtracting 1 from each of the digits in 365. Additionally, the digits in 254 add up to a sum of 11. Since the 9/11/2001 attack on the *twin* World Trade Center towers was the *second* attack on these structures, we can multiply the date sum of 11 by 2 and again get 22!

Should the reader require additional evidence that there is a connection between numerical data derived from the pyramid in the Great Seal of the United States and the timing of terrorist attacks on the United States, then consider this. In the second numerological analysis I provided for the image of the engraved pyramid on the back of the US one dollar bill earlier in this chapter, I directly measured the width of the seventh course of stones and found it to be about 9 millimeters. A measurement of the pyramid's height yielded a figure of about 11 millimeters. From these measurements we can directly obtain the month and date of the second attack on the World Trade Center towers which were the 9th month and 11th day!

Of course, all of these numerological analyses of the dates of terrorist attacks on the United States could, if taken individually, be due to coincidence. However, when they are considered together, they would seem to more likely indicate some sort of intentional timing by the terrorists that cannot be attributed to mere chance. The terrorists must have selected these dates for a reason and it would seem suggested by the preceding material that it could be related to the number 222 embedded in the structure of the pyramid on the reverse side of the Great Seal of the United States of America or directly to the dimensions of the engraved pyramid itself.

Finally, as all American citizens know, our Treasury Department fairly recently began changing the design of our paper currency. This change was necessary to incorporate the latest security safeguards into the currency to help minimize its counterfeiting around the globe. I will not delve into the exact safeguards now appearing in the paper currency of the United States except to note that they are very elaborate and difficult to duplicate. As of the date that this chapter was completed, the 100, 50, 20, 10, and 5 dollar denominations have been changed over to the new design and there are plans to reintroduce a 2 dollar note with the new format soon. So far, the changes from the previous designs have been modest and people seem to have gotten used to the new look of our currency. However, I am wondering if the conversion will continue right down to the one dollar bill since this small denomination is rarely counterfeited. If it does, then I am wondering whether or not *both* sides of the Great Seal will still be present on the back of the note and, if they are, how the pyramid will be depicted. Will its engraving be further obscured so as to make extracting numerological data from it virtually impossible? Or, will the pyramid be removed completely?

Complete removal of the pyramid might be perceived by terrorist organizations as a sign that the promised protection given the United States of America by the Washington Prophecy was no longer in operation. Thus, removing the pyramid from the back of the one dollar bill might embolden the terrorists to continue their attacks against the United States, both domestically and abroad, by making them think that they were more likely to achieve victory against America now that its divine protection was gone. My suggestion to the United States Treasury would be to retain both sides of the Great Seal on the back side of any new design for the one dollar bill and to make the engraving of the pyramid's 13 courses of masonry even more precise so that the individual blocks of stone are easily discernible from each other within a course. This should send a message to the terrorists that continued attacks against the United States of America, in particular, and the west, in general, will be futile.

I, for one, will be keeping the number 222 and the fate of the Great Seal of the United States in mind as I watch the war on terror and terrorists play out in the years to come.

Chapter 6

The Coming End of Fossil Fuels

As we begin the present 21st century and 3rd millennium, the world finds itself in somewhat of a predicament. Mainly, we have inherited a culture that is the offspring of an Industrial Revolution which has made us highly dependent upon the combustion of fossil fuels to meet most of our current energy needs. It is the energy from this combustion that spins the generators whose electrical output lights our homes and powers our factories. It propels our automobiles, commercial aircraft, and military hardware.

It was only fairly recently thought that the global supply of fossil fuels was limited with the most pessimistic estimates suggesting that all currently known and projected coal, petroleum, and gas deposits would be completely gone in another ten to twenty years. Surprisingly, this prediction has been completely reversed! Due to advances in extraction technologies, it seems that there is now an abundance of fossil fuels available to humanity and new estimates see projected supplies lasting for hundreds, perhaps, even thousands of years into the future.

While this might seem like a desirable situation, there is a very serious problem with it. It is impossible to extract the chemical energy of fossil fuels without also creating huge volumes of carbon dioxide which are then released into our planet's atmosphere. The molecules of this gas, one of several "greenhouse gases", then act like submicroscopic antennas which are capable of absorbing infrared radiation from our Sun as it enters Earth's lower atmosphere. Upon absorption, the molecules become highly agitated as they vibrate and spin about. It is not long before their enhanced kinetic energy is then transferred to other surrounding atmospheric gases (mainly nitrogen and oxygen) with the result that the entire atmosphere begins to slowly heat up during the daylight hours.

With a slowly rising average atmospheric temperature for our planet, other processes then become triggered. For example, the warmer air expands a bit and can physically hold more water in vapor form. This can result in worsening droughts and falling ground water levels in regions that depend upon adequate water supplies for agriculture. Ocean water temperatures also rise and, aside from interfering with the lifecycles of various sensitive marine organisms, the warmer waters circulating toward the Earth's polar regions can cause accelerated melting of glaciers there and the disappearance of sea ice. Inevitably, the extra melted water raises the ocean's water levels and can inundate coastal regions and make them uninhabitable for our planet's growing global population.

The truly frightening thing about these effects is that they can, in time, become "cascading" or runaway ones. That is, as they start out slowly and then begin to gradually worsen, they can begin a phase of uncontrolled acceleration in which the problems escalate to the point where they actually become irreversible through any human efforts and, eventually, become so bad that they could threaten the lives of a majority of the lifeforms, including humans, which inhabit the Earth. Currently, there are some predicting that this problem of runaway "climate change" (formerly referred to as "global warming") is *already* at the point of no return and will be getting catastrophically worse as the years pass and the green house gas pollution from the ongoing industrialization of various economically "emerging" nations is added to our atmosphere.

One might think that, with such dire forecasts, the entire planet would be uniting in an effort to stop the process by immediately implementing alternative methods for generating "clean" energy that do not rely upon fossil fuel combustion and the atmospheric release of additional amounts of carbon dioxide. Yet the sad reality is that little is being done about the problem on a global level. (Another major greenhouse gas is methane. It is mainly emitted from the digestive tracts of large herds of farm animals raised for their meat as they consume and digest the carbohydrates found in the various grasses upon which they graze. This flammable gas is also emitted in large quantities during the extraction of underground supplies unless the wells are properly sealed at the time that they are drilled.)

The reason that the world currently seems addicted to using the various carbon rich fossil fuels is simply that researching and then implementing new cleaner technologies on a wide scale is an expensive process while the most financially profitable process is to just keep using currently extracted fossil fuels and to do so as long as the newly discovered abundant sources last. Those that advocate the latter approach tend to either deny the reality of or minimize the problem of the currently escalating climate change. If they do acknowledge its reality, they may then just simply dismiss it as being part of some natural process that the Earth is currently going through that has little to do with human created atmospheric pollution and which, given enough time, our planet will somehow correct on its own.

This optimistic approach is currently the one that the majority of Earth's nations are pursuing especially the poorer or "Third World" nations whose populations are trying to climb out of poverty through the development of their local economies, particularly their industrial sectors which tend to use the cheapest fossil fuel provided power and release the most carbon into the atmosphere. As an example, consider that even mainland China has an ongoing and unprecedented upsurge in industrial production and automobile usage. With these comes a greater thirst for the energy that was slowly stored up and locked into the Earth's crust many millions of years ago and they are now building a coal fueled electric power plant at the rate of about one per *week*!

The situation in the United States has, to a certain extent, been exacerbated by the American motorist's currently fashionable desire to own and use a conveyance known as a "sport utility vehicle" or "SUV". These are large, bulky, high center of gravity versions of the old station wagon and only average about 15 miles per gallon of gasoline burned. It seems that before too long the SUV will account for about half of all of the vehicles owned in the United States. Because of the higher profit per vehicle for these, the automobile industry here is in no hurry to limit their production.

In an effort to buy some time, the United States government keeps the automobile industry under pressure to come up with ways to boost the mileage per gallon of gasoline used in its products. Thus, we are beginning to hear talk about such things as "hybrid" drive and fuel cell powered vehicles. These are, at best, only short term solutions to try to slow the rate at which carbon is being added to our planet's atmosphere. They are still expensive technologies that will add to the initial cost of a vehicle and its later maintenance and, thereby, reduce any savings in the cost of fuel that they might produce.

Hybrid automobiles are currently being manufactured by several Japanese automakers and are beginning to arrive in the US market in increasing numbers. The basic concept is that the car incorporating such a hybrid system is powered solely by an electric motor during the times when its fuel powered engine tends to waste fuel which is usually during the brief intervals when the vehicle is accelerating. Thus, during acceleration the car is electrically driven by its electrical motor and then switches over to being driven by its gasoline engine while it is cruising at a more or less constant speed. When cruising, the gasoline engine also recharges the car's batteries which were partially drained when its electric motor was used during acceleration. Alternately using a gasoline engine and an electric motor for a vehicle's propulsion allows one to approximately double the mileage per gallon of gasoline burned. This, of course, means that these hybrids only add about half as much carbon to the atmosphere per mile driven as compared to that added by a solely gasoline engine powered automobile.

Currently, hybrid cars cost thousands of US dollars more than conventional gasoline engine only powered vehicles and the consumer must hope that he will recover the extra initial expense over the life of the automobile as he saves some money at each trip to the gas station. Whether these hybrid vehicles are truly "cost efficient" in the long run still remains to be seen as they are only beginning to appear on American roads in relatively small numbers. Their main obvious disadvantages are the extra and expensive batteries they carry which must be replaced every few years and the complex, computerized control systems required to allow them to switch back and forth smoothly between both types of drive.

Surprisingly, there have been several inventions during the past century which produced even greater increases in gasoline mileage than the newly arrived hybrid systems and which were far simpler in construction and, if used now, would only produce a negligible increase in initial vehicle cost. I recall that during the 1960's I read advertisements for an ingenious device called the "Porg carburetor".

A conventional automobile gasoline engine using this novel carburetor was started as usual. After a few minutes of operation, the engine's exhaust manifold would become hot and this heat would then heat up a special chamber which was mechanically attached to the manifold. After a few minutes of driving, one would then pull a handle on the dashboard that was connected, via a cable, to a switching valve on the engine's fuel line. The valve, once activated, would then divert the flow of gasoline from the engine's carburetor and into the heated chamber on the exhaust manifold. Once inside the chamber, the gasoline would rapidly boil to form a pure gasoline *vapor* which would then leave the chamber and travel through a heated fuel line to finally enter the carburetor for use by the engine. At this point, the engine would begin running solely on a mixture of air and heated gasoline vapor. Apparently, tests showed that, even with a mid sized car, this carburetor was capable of delivering better mileage per gallon of gasoline burned than is currently being done with fuel injected engines.

As the reader interested in automotive engineering will realize, the main problem with gasoline engines that significantly cuts down on the mileage they produce per gallon of fuel they consume, is the physical size of the gasoline mist particles drawn into their cylinders for combustion. A cold, standard carbureted engine produces the worse and lowest mileage because it draws in the largest size gasoline mist particles. These tiny fuel droplets tend to burn on their outer surfaces only and are not completely burned up before the contents of a cylinder are expelled during the "exhaust" phase of a cylinder's four stroke cycle. Thus, one can often smell raw gasoline in the exhaust of such an engine.

The next best improvement in gasoline mileage was achieved with the use of a "fuel injected" engine. This uses a special carburetor containing one or more injector nozzles that literally spray a very fine mist of gasoline droplets into the air stream drawn into the engine's intake manifold by the various cylinders which at any moment are in the "intake" phase of their four stroke cycle. Smaller mist particles burn more thoroughly to extract more explosive energy from each gallon of gasoline burned.

The fuel injected engine was then further improved by using what is known as a "rail injector". This is simply a high pressure metal pipe that is routed around the engine's cylinder blocks and which provides pressurized gasoline to *individual* injector nozzles that spray gasoline mist particles *directly* into each of the engine's cylinders. Because the mist particles spend slightly less time in the air before they are ignited, they are less likely to recombine to form larger, less efficiently burning particles as they would if they had to travel through a long intake manifold before reaching the engine's cylinders.

All of these engine modifications will improve gasoline mileage by reducing the mist particle size of the fuel, but they can not compare to the use by an engine of a pure fuel vapor. In such a vapor there are no droplets and the mist actually consists of individual fuel molecules. Combustion of such a vapor will be close to 100% efficient. The Porg carburetor, by using only a pure fuel vapor once the engine had reached operating temperature, apparently delivered such performance (one test I remember reading stated that it was possible to get 75 miles per gallon with a Ford "Galaxy" automobile which is a fairly heavy midsized vehicle), yet I have heard little mention of this invention as a viable solution to reducing the amount of gasoline burned per mile driven and, in the process, reducing the amount of carbon emitted into our atmosphere per automobile per year. I often wonder why this is so.

I do not want to entertain various conspiracy theories in this chapter about how groundbreaking and potentially oil company profit threatening inventions are suppressed. However, I will simply state that if the automotive industry is serious about improving gasoline mileage and thereby reducing the amount of carbon emitted from their vehicles, then they should be considering the Porg carburetor (and other variations of it) for use in automobiles. These devices are simple and inexpensive to construct and add to a vehicle and have virtually no maintenance costs. Unlike the presently touted hybrid vehicles, automobiles with these specialized carburetors do not require extra, heavy batteries or complexly programmed microprocessors for operation. And, as an added benefit, because they use pure fuel vapors for combustion, that combustion is probably as "clean" as can be practically achieved while using a hydrocarbon fuel. The exhaust gases that cars using these device's produce will consist almost entirely of carbon dioxide and water vapor with little, if any, carbon monoxide and practically no unburned, raw gasoline.

One of the reasons that we now see hybrid vehicles being provided by the Japanese auto industry and only very limited production of such vehicles by the American car makers is because the American auto industry is currently spending considerable amounts of research dollars attempting to perfect all electric cars that will be powered solely by fuel cells. Fuel cells have been around since the early 1960's and are practical and very efficient devices for producing electrical power. They can be made to provide an external electrical current from virtually any chemical reaction wherein electrons are transferred between two chemical reactants. The two reactants which provide the maximum electrical power output from a fuel cell are the simple gases of hydrogen and oxygen and fuel cells using these gases have been used on spacecraft for decades. After producing its electrical power when these gases are fed into it, the fuel cell's only "exhaust" is harmless water which goes off into our atmosphere without causing any environmental damage or contributing to the so-called atmospheric green house effect.

I have been hearing about the possible use of fuel cells in automobiles for decades and was happy to hear about their imminent usage in American automobiles before the year 2010. My joy was short lived, however, when I began to learn the details of this "new" approach to powering our automobiles.

The problem with hydrogen / oxygen fuel cell powered automobiles is that, while there is plenty of atmospheric oxygen for them to use, they must still be provided with a supply of hydrogen gas. This hydrogen gas must be obtained from some chemical substance by using electrical power to extract it. Early pioneers in the development of fuel cells envisioned the hydrogen they needed being extracted from sea water via the use of solar cell generated electricity. Thus, it was originally believed that fuel cells would mark the beginning of a 100% pollution free, environmentally friendly, low cost solution to humanity's future energy needs. The actual plans for this remarkable technology are, sadly, beginning to deviate from the futurist's optimistic vision for it.

Already, it is becoming apparent that the hydrogen for our imminent automotive fuel cell revolution will not be derived from sea water. The reason is because, when extracting hydrogen from water, one must break a chemical bond between an oxygen and a hydrogen atom and this bond is one of the stronger ones in nature and, thus, requires more electrical power be used than when a bond is broken as a hydrogen atom is pulled away from an atom *other* than oxygen.

Next to being bonded to an oxygen atom in nature, a hydrogen atom is most abundantly found bonded to an atom of carbon and is bonded more weakly to this atom than to one of oxygen. Because of this, it has now been more or less decided that the hydrogen atoms needed for the "hydrogen revolution" will not be derived from water, but from the long chain carbon molecules found in *petroleum*! If this approach works, then, via the application of electrical power, the hydrogen atoms can be removed from the petroleum without sending its carbon atoms off into the atmosphere as parts of carbon dioxide molecules. The only product of the extraction should, under ideal circumstances, be hydrogen gas and a dark, carbon laden slag. So far, I have not found much mention about what will be done with the slag left over after the hydrogen atoms are extracted from the petroleum. Hopefully, it can be buried and will harmlessly degrade somehow without causing any serious environmental damage.

From this we see that the proposed use of fuel cells for automobiles will require both a plentiful source of hydrogen gas *and* electrical power for the extraction process and it is this second component that may render a global "hydrogen economy" ultimately unfeasible. This is because most of the world's electrical power is still being generated by coal burning power plants which are responsible for adding the most carbon to Earth's atmosphere. The alternative to this is to use natural gas as a fuel for the power plants. The cost of natural gas has, only recently, come down somewhat because new extraction technologies have increased the supply. But, it still requires a considerable investment to convert a coal burning power plant into a gas burning one and this cost is inevitably passed on to the consumer.

One hoped for alternative to either the use of coal or gas to make possible the proposed pollutionless hydrogen economy of the future is to simply increase the use of other, non polluting power generation technologies. Unfortunately, currently only a small fraction of humanity's electrical power needs are being provided by such alternative sources as nuclear, hydroelectric, wind, geothermal, and solar electrical power generation. Each of these alternatives to fossil fuel combustion generated electricity has its limitations and drawbacks. Let's very briefly consider the problems with them.

Once considered the ultimate solution to mankind's energy needs, nuclear power is actually far more expensive and dangerous than the use of fossil fuels! Consider that a single troy ounce of plutonium for a typical reactor's core now costs tens of times more than the same weight of gold and a nuclear power plant reactor can require tens of *tons* of this fissionable material. In the process of extracting the nuclear energy from the plutonium in the reactor core's fuel rods, each fuel rod becomes a *million* times more radioactive than when it was first placed into the reactor. A fuel rod that was safe to hold in one's hands before it is used eventually becomes so radioactive after a few months of being in an active reactor that it would kill a person with radiation poisoning who was standing next to it without protective garments in less than a minute!

Plutonium is also considered one of the most toxic materials ever isolated on Earth and a single microgram of the metal, if inhaled, will produce fatal lung cancer in about a two week period. With all of these negatives, it is any wonder that people do not want nuclear power plants built anywhere near them. In the event that a reactor's cooling system should fail, there is the possibility that its very heavy and highly radioactive core of fuel rods could overheat and become a molten mass that would melt right through the bottom of the steel reactor vessel which contains it. It would not stop there, however, but would continue melting through the concrete floor of the containment building until it reached the water table below the reactor building. The water there would soon begin to boil violently and a cloud of radioactive steam would fill and pressurize the dome of the concrete containment building. Should that dome fail and be blown open by the growing pressure, then a variety of radioactive gases (including plutonium vapors) would be released and carried along by prevailing winds. The net result is that such an unfortunate accident could contaminate an area of land the size of a state!

It is primarily for these reasons that nuclear power plants are no longer being constructed in the United States. Although the technology does work, it has potentially catastrophic dangers associated with it. Even if a nuclear power plant functions flawlessly, there is still the problem of what to do with all of the spent and highly radioactive fuel rods that have to be periodically removed from its core. So far, no long term solution has been found for the disposal of these rods which are currently kept on the premises of the nuclear power plants in the US. Most people do not realize that, in operation, nuclear power plants actually *create* highly radioactive materials some of which will remain dangerous for millions of years! If nuclear power is selected to provide the electrical power required to bring about a global hydrogen economy, then I can see only problems coming from this decision in the future. The only bright spot in the nuclear option picture is that there are currently new and improved reactor designs on the drawing boards that, apparently, can not experience a meltdown and release of radioisotopes in the event that their cooling systems fail. These new designs, however, still produce the same amounts of nuclear waste as do the older reactors still in use.

I won't spend much time here discussing another envisioned energy generation technology known as "fusion power". Its goal requires forcing the nuclei of hydrogen atoms to "fuse" together to form helium nuclei and, in the process, release abundant energy. This is possible because the resulting helium nucleus then has slightly less mass than the two hydrogen nuclei from which it was formed and this mass deficit then becomes pure electromagnetic energy that is then emitted in the form of a high energy gamma radiation photon as the fusion takes place.

The gamma radiation that would be emitted as trillions of helium nuclei form via fusion per second could then be used to heat water until it became steam which then would drive a steam turbine that would turn a conventional electrical generator to produce electricity. It all certainly sounds fairly easy to do, but the problem is that two hydrogen nuclei that must be made to collide with each other each carry a positive electrical charge that makes them strongly repel each other as one attempts to force them together. From studying the mathematics of this situation, one quickly realizes that one must be able to artificially create conditions of extreme temperature and pressure in order to cause fusion to take place. Ordinarily, these conditions only exist in such things as stars and exploding hydrogen bombs and all attempts, so far, to produce them in the lab have been disappointing. This situation is likely to remain so for the foreseeable future unless there is a revolutionary breakthrough in the design of the "magnetic bottles" used to contain hydrogen nuclei as they are energized in order to make them undergo a steady fusion reaction. Thus, we probably will not be able to rely on fusion power to provide the electricity needed to make an envisioned global hydrogen economy work at the present time.

Hydroelectric, wind, and geothermal power plants are now providing clean electrical power that becomes more "cost efficient" as they continue to operate. Unfortunately, they require that special geographical, geological, and meteorological conditions exist in order to make them possible. Currently, they are only providing a few percent of the world's energy needs and are insufficient for the needs of a global hydrogen economy.

Finally, we come to the matter of solar power generation. This is just natural fusion power as opposed to the artificial version produced in hydrogen bombs or hoped to eventually be produced in manmade fusion reactors. When using it, we let the Sun to do its normal job of fusing hydrogen nuclei in its core and then we simply intercept some of the resulting radiant energy that reaches the Earth and turn it into electricity with large arrays of photovoltaic cells. Again, this technology does work and is providing some electrical power for satellites, space stations, and homeowners trying to lower the cost of their electric bills and generate less air pollution. Unfortunately, even the latest and most advanced "solar panels" only convert a small percentage (less than 10%) of the radiant energy they intercept into electrical energy. These panels are expensive to make and necessitate the use of exotic materials that require considerable energy expenditure in order to obtain and refine. Once again, I do not foresee any sort of large scale use of solar energy to make the electricity available that we will need to convert over to a global hydrogen economy at the present time. But, once again, this assessment could change if there is a revolutionary breakthrough in the power output of solar panels.

So, if we can not risk the continued atmospheric pollution from the combustion of fossil fuels or the currently available, but expensive, environmentally "friendly" electrical energy production technologies, then what other possible energy source could be used to finally make a global hydrogen economy possible? Actually, if we could generate enough electricity, then we would only need the hydrogen extracted with it from water molecules for fuel cell use in places where it is not feasible to run electrical power lines such as to automobiles, aircraft, and remotely located dwellings. So, the ultimate problem that humanity must deal with in the coming century is the search for some technology that will allow for the virtually unlimited generation of electrical power without the undesirable side effects of fossil fuel's atmospheric carbon pollution and nuclear power's toxic and highly radioactive wastes.

After giving the matter much thought, I have reached the conclusion that the ultimate solution we seek will eventually come from the field of *gravity* physics. From my study of the extraterrestrial UFO phenomenon, it has long been apparent to me that whoever is constructing and operating the UFO's that have been sighted in our skies for the last several thousands of years has managed to develop a technology that is capable of rendering normally massive objects either partially or completely massless.

Thus, a large diameter (say 32 foot) metallic disc shaped craft that would normally weigh in the tens of tons can, via the technology used by the builders of the UFO's, be artificially rendered massless and, therefore, weightless and inertialess for various periods of time. I believe that this is achieved by a device that I refer to as an "anti-mass field generator" in my writings on the subject. I can not go into all of the details of these remarkable hypothesized devices in this chapter (the curious reader is referred to the author's previous three volumes on UFOs for a more thorough treatment of anti-mass field generators) except to say that they contain a rotating magnetic field that is made to move along its field lines at right angles to an electric field. The result of this action is that a new form of *non*-electromagnetic particle radiation which I call "anti-mass field radiation" is then emitted from the anti-mass field generator.

In a UFO, the emission of its anti-mass field radiation serves to neutralize or cancel out the normally emitted "mass field radiation" that issues from every mass possessing subatomic particle of the structural atoms of which it is composed (and those of its crew if it carries one). Once the intensity of the anti-mass field radiation emitted from a UFO's anti-mass field generator equals the intensity of the mass field radiation that the craft (and its crew) emits, then the UFO (and its crew) becomes *completely* massless, weightless, and inertialess. Should the intensity of the anti-mass field radiation emitted from a UFO exceed the intensity of the emission of mass field radiation from it and its contents, then there is *excess* anti-mass field radiation present which can begin to reduce the mass of materials near the craft's hull such as the layer of air immediately surrounding it. It is this emission of excess anti-mass field radiation which accounts for practically all of the strange effects reported in the UFO cases.

Now imagine what the ability to render a heavy weight instantly weightless could mean for the production of energy. Let us consider a machine shaped like a Ferris wheel that is, perhaps, 50 feet in diameter.

Instead of cars to carry passengers, we attach heavy metal weights to the frame of the wheel. Assume that each weight is a sphere of pure lead that normally weighs one ton which is equivalent to 2000 pounds. Further assume that there are 8 of these weights that are arranged at 45 degree angular intervals around the circumference of the Ferris wheel shaped machine.

Now the wheel portion of the above design will carry a total of 8 tons of lead weights and will have no tendency to rotate about the axle that connects it to its support structure. This is because the gravitational pulls on both sides of the Ferris wheel are always equal since the quantities of lead on both sides of the wheel are always equal. Further assume that we have attached a small electrical motor to the axle of the Ferris wheel which we can use to rotate the massive wheel at a low speed. If we used the motor to slowly turn the wheel and then turned off the motor, we would find that the wheel would slowly come to a standstill due to atmospheric drag and any small amount of friction present in the small bearing of the motor and the larger bearings on which the axle rotates.

Let us again consider the above described Ferris wheel-like machine and its lead weights, only this time we will make a modification to the design. We will equip *each* of the lead weights with its own small anti-mass field generator which can instantly (when it is activated) reduce the mass and weight of its sphere of lead to zero during any time interval that we choose. Using a simple optical control system, we arrange for all of the lead weights on only *one* side of the Ferris wheel's axle to become massless when they are on that side. Thus, we arrange matters so that, should the wheel be allowed to rotate about its axle, whenever a lead weight with its normal mass and weight of one ton moves around the *bottom* of the wheel to that one side of the wheel its particular anti-mass field generator will be switched on and the weight will immediately become massless and weightless. However, whenever during wheel rotation a massless and weightless weight *leaves* that one side of the wheel and then travels over the *top* of the wheel to the other side, its anti-mass field generator is switched off again so that the weight immediately regains its normal mass and weight of one ton.

Once this modification in the design is made, we will find that a remarkable effect is taking place. The wheel portion of our machine is now in a chronic state of imbalance and, if allowed to, it will begin to rotate rather robustly. The side of the wheel on which the lead weights have their normal mass and weight of one ton will rapidly rotate toward the ground, while the side of the wheel carrying the massless and weightless lead weights will rapidly rotate upward toward the sky. As massive lead weights on the *descending* side of the wheel pass through "bottom dead center" in their movement around the wheel and suddenly find themselves on the ascending side of the wheel, their anti-mass field generators will be switched on and they will instantly become massless. Also, as the massless lead weights on the *ascending* side pass "top dead center" in their journey around the wheel, they will pass over onto the descending side of the wheel at which time their individual anti-mass field generators will be switched off and they will then instantly *regain* their normal mass and weight.

We will find that, as long as the anti-mass field generators are allowed to only emit their mass canceling anti-mass field radiation on the ascending side of the wheel, the entire wheel will rotate with considerable torque. As the lead spheres drop on the descending side of the wheel, they will lose a tiny portion of their "rest mass" which will be converted into the kinetic energy that rotates the wheel. However, as these same lead spheres rise on the ascending side of the wheel, they do so while massless. Thus, although the lead spheres rise in Earth's gravity field on the ascending side of the wheel, they do *not* regain any of their lost rest mass and no rotational kinetic energy is removed from the wheel in the process! In effect, as long as the wheel of our machine is allowed to rotate, it will continue to convert the mass of its lead weights into pure mechanical energy with 100% efficiency!

As the wheel rotates on its own, the small motor that we originally attached to it can now be driven in reverse by the torque of the wheel and should then function as an electrical *generator*. A small percentage of its power output will, however, have to be fed back into the rotating wheel (via some sort of arrangement of slip ring electrical contacts) in order to provide electrical power to its eight anti-mass field generators (four of which will be active at any moment). But the remainder of the electrical power provided by the motor now functioning as a generator can be considered to be "free" energy which is produced without the use of conventional fuel and without the emission of any atmospheric pollution or radiation.

By using several such Ferris wheel-like machines together it should be an easy matter to drive a bank of standard alternating current generators that would then collectively feed their outputs directly into the electrical grid of a country. For small countries which do not possess a sophisticated electrical distribution system, each individual town or village could be directly supplied with all of the electrical power it needs by a small "wheel farm" of such machines located outside of the town.

One wonders what will eventually happen to the lead weights we have used in the above envisioned Ferris wheel-like machine as they move repeatedly around its axle and thereby continue to have the energy associated with their normal rest masses extracted from them. The most obvious scenario is that the weights would just continue to lose mass until a point was reached where they became completely massless and weightless. At this point the wheel could deliver no more mechanical energy to the generator attached to it and electrical power generation would cease. For wheels whose weights experience this fate which were being used for commercial power generation, the weights would probably be replaced after they had lost only a few percent of their normal masses and weights. If the mass reduced weights removed from these wheels were then allowed to "rest" for a while outside of a wheel, then perhaps they would somehow recover their lost rest masses by some mechanism.

However, there may be several alternative fates that can befall the weights used inside of an active wheel generating electrical power.

Instead of the mass loss being evenly spread amongst all of the mass possessing subatomic particles of the lead atoms of a weight as mentioned above, perhaps just *individual* whole lead atoms within each lead sphere will begin to randomly and completely disappear so that, over time, the density of the weights will still slowly diminish. If this scenario occurs, then a point may be reached where the weights will have an internal microscopic structure riddled with holes once occupied by lead atoms and will then be too fragile to be of further use for power generation. If this proves to be the case, then the used lead weights can be removed for recycling (i.e., melting down in order to cast new, more solid weights) and immediately replaced with new weights having their full mass and weight so that power generation can continue.

However, it is also possible that only individual single pairs of electrons and protons or single neutrons will begin to randomly disappear from within individual lead atoms of a weight. Should this occur, then these atoms' nuclei will begin to become unstable and undergo nuclear transformations. This is the least desirable scenario because there is the chance that the lead atoms involved will be converted into *radioactive* isotopes of various elements lighter than lead. If this effect does occur, then this proposed energy generation technology may not offer any long term safety advantages over today's nuclear power. Used weights from the Ferris wheel-like electrical power generators would then, once they reached a certain degree of radioactivity, have to be removed and stored securely, possibly for millennia, until they no longer posed a radiological hazard and they would then be replaced with fresh weights containing only nonradioactive lead atoms again so that electrical power generation could continue.

There is also one finally and rather intriguing possibility which is that the lead spheres used will not experience any slow diminution in mass and can, therefore, be used indefinitely. If this proves to be the case, then one might legitimately ask how our Ferris wheel-like machine obtains the mechanical energy that it uses to generate electricity. On the surface this possibility would seem to be an obvious violation of the "First Law of Thermodynamics" which is a physical law that has never been known to be violated.

I am, of course, a firm believer in the inviolability of the famous energy conservation law, so if the weights of such a machine do not lose mass when it has been in operation for a while, then it could only mean one thing. It would imply that each of the massive lead weights actually was losing a small amount of its rest mass during each rotation of the wheel which carried it, but that mass was then somehow *immediately* restored to the weight by energy taken from its environment! Perhaps as a small amount of mass was continuously restored to a weight during each second of wheel operation, there would be a continuous drop in the weight's temperature to "pay" for this mass restoration. Or, perhaps, the energy needed to restore the mass of a weight would come from our Earth's molten interior via the ambient magnetic field that it produces and which, locally, penetrates the weights of the Ferris wheel-like machine.

Eventually, these matters will all be resolved. The important thing now is that serious research into the mass altering effects displayed by UFOs begins to take place. If UFOs are real as I and many others firmly believe, then this novel method of electrical power generation should also be possible. After studying the UFO phenomenon for decades, I have no doubt that the ability to artificially control mass (and the properties of weight and inertia associated with it) will not only allow us to build our own operational UFO-like craft, but will also allow us to supply all of the pollutionless electrical energy needs of humanity well into the future.

Chapter 7

Thoughts on Permanent Magnet Motors

We LIVE IN A WORLD which, year by year, is becoming increasingly concerned about where "clean" energy to power our global civilization will come from as we proceed through the 21st century. The overly optimistic promises of fission and fusion power that were made in the previous century have not been fully kept and we have been mostly left dependent upon a supply of fossil fuels whose toxic combustion products are slowly damaging our planet's atmosphere and altering her weather. If ever there was time for a revolution in energy generation to take place, it is now. Yet, the alternatives to fossil fuels have all proven to be plagued with various technical problems.

Alternative energy generation technologies have been shown to work, but are, generally, too expensive and temperamental to meet the growing needs of the world's energy hungry peoples. Conservation programs and improvements in the efficiency of energy usage have helped a little. At least they have slowed the rate of increase in fossil fuel usage somewhat. But, in the final analysis, if humanity is going to survive ever worsening climatic problems and continue to progress economically, then it is imperative that we find some vast new supply of energy which can be abundantly and easily tapped.

For many years, especially in my youth, I realized that the area of alternative science that dealt with energy production might provide that vast new supply of energy mankind would eventually need to harness. Indeed, I have devoted much time in my life to a search for a device that might be able to deliver virtually unlimited amounts of energy. In pursuit of this end, I have spent many hours contemplating, designing, and constructing various devices that I had hoped would be able to provide a steady and reliable supply of what is often referred to as "free" energy since it apparently does not have a conventional source. My efforts were mostly directed toward the construction of rotating wheels that would use shifting weights within their interiors to provide a continuous torque that could then be used to perform external mechanical work such as driving electrical generators to obtain power in the more convenient form of electricity. Unfortunately, all of my efforts were futile although I remain, to this day, convinced that it can and has been done several times in the history of this planet.

As I studied the topic of "self motive" power in depth, I became aware that the devices intended to achieve this "Holy Grail" of science and technology fell into several distinctive categories.

Firstly and mainly, there were the mechanical devices like the ones I was pursuing. They utilized various arrangements of weights, springs, and levers that worked with gravity in an attempt to produce more energy in operation then they consumed in order to stay in motion. These were followed by machines that it was hoped could use the differences in air or water pressure to provide energy most of which would be outputted to the environment to perform useful work while the device that produced this free energy would always retain a smaller portion of it in order to continuously restore itself to some initial state. Thus, these devices were intended to operate in a closed cycle that would repeat itself indefinitely.

Moving away from gravity, one finds devices in which the attractive and repulsive forces acting between permanent magnets were intended to maintain the rotation of a shaft, the oscillation of a pendulum, or the rocking of a beam all of which motions could be used to turn a generator's shaft and produce electrical power. Because magnetic fields tend to be stronger than gravitational fields, these devices promised to produce more powerful torques with a more compact design than could a gravity activated device. Moreover, unlike a device dependent upon maintaining a fixed orientation with respect to the Earth's gravity field, a magnetic device should be able to operate regardless of its orientation with respect to our gravity field.

Devices that relied on other than gravitational or magnetic fields were also possible and, occasionally, one would find a design that used simple electric fields to keep a wheel in rotation. Early electrical experiments in the 18th century resulted in several devices (consider Benjamin Franklin's "Electric Wheel") which would rotate mysteriously for long periods of time although usually with low torques. But, the electric charge differences in the components of these devices that were responsible for the electrostatic forces that then resulted in the motion of their parts eventually dissipated and they all came to a stop. None of them was ever able to use part (or even all) of its axle's torque to *fully* electrically recharge itself while it ran so that it could do so indefinitely.

Finally, there was a small category of devices that first appeared in the late nineteenth and early twentieth century and that involved *current* carrying electrical or electronic circuitry. Like those cited above that used pure electric fields for their motion, these devices were intended to produce electrical power continuously in excess of a small amount of power that they consumed while in operation. Often these required a battery to start themselves up after which they would produce electrical current, usually via some sort of electromagnetic induction process, that could then be outputted to power an external electrical motor or heating element while a part of the produced current recharged their starting battery. Ideally, after the battery was fully recharged, it could be disconnected from the circuitry and the device would become perpetually self powered. None of them ever did, however.

As I and others have noted, the pursuit of these devices seemed to mirror the development of the various engineering sciences and some even say that it was actually the pursuit of such exotic devices which then spurred the development of the various fields of engineering!

In my early research into this subject, I did make several occasional attempts at building devices which used permanent magnets to produce continuous power. I did this despite the fact that all of my science teachers considered it an impossible feat to accomplish. Their reasoning certainly made intellectual sense to me, but somehow I felt at the time that there might be some unique arrangement of magnets that would allow me to overcome their objections. At the time I was unaware of just how many individual inventors during the last several millennia had the exact same intention as I did.

Over the years that my serious involvement in the pursuit of a genuine free energy device spanned, I managed to make the acquaintance of several individuals who, like me, tried inventing various similar devices as a hobby. One day in the early 1980's, I received a telephone call from one of them who was a much older gentleman than I whom I shall hereafter refer to by the pseudonym of "Bernard".

When he called me, Bernard was quite excited and told me that while he had been perusing a recent issue of a popular magazine of interest to inventors, he came across an advertisement in its classified section that offered plans for sale that would enable one to construct a *genuine* free energy device. He was surprised that I had not heard of the new device especially since it was patented!

My curiosity was immediately aroused about the invention, especially since it claimed to produce free energy and had also been granted a patent. The United States Patent Office had stopped issuing such patents at the beginning of the 20th century on the grounds that such a device was physically impossible and they did not want to waste time processing their patent applications which would inevitably be rejected. However, there was *one* exception to this rule. The Patent Office would, indeed, grant such a patent if the inventor possessed and could demonstrate to them a *working* physical model of the device. I told Bernard all of this and I could feel his excitement growing even stronger over the phone. At this point we were both convinced that the long search for such a device had finally ended and that a Golden Age of unlimited free power production was about the begin!

Bernard, who lived modestly on a monthly Social Security check and the small interest on a retirement savings bank account, decided to order the plans at once despite the fact that they cost about $50. He was eager to build a copy of the invention that he could then proudly display to all of the people who had scoffed at his own attempts to construct a free energy or "overunity" energy machine during his lifetime ("overunity" meaning that the device created more energy than it used to keep itself running and, thus, produced some extra "free" energy that was then available to perform work external to itself). He even hoped that he might be able to somehow improve upon it and get his own patents on the improvements.

About three weeks after our initial conversation about the device, I received another call from Bernard. He had received the plans for the invention and it turned out to be a kind of motor that was powered by permanent magnets. Aside from some of the odd shaped magnets it used, its construction was fairly simple. Bernard, however, wanted to make an *exact* copy of the device and intended to use the services of a local machine shop. They would fabricate all of the parts according to the blue prints Bernard had received and then he would have the pleasure of the final assembly of the motor. He planned to use it to run a small AC generator whose power he would then use to operate a bank of light bulbs. It would be an impressive demonstration of the new technology that would delight all to whom he showed the device.

Copying this motor, however, turned out to be a lot more difficult than either Bernard or I anticipated. The machine shop charged him close to $1500 to make the various metallic parts of the motor which were carefully and precisely cut from standard steel and aluminum stock. But he considered it a small price to pay in order to finally be able to prove to himself and his family that what he had pursued for several decades was real.

Finally, after several weeks, the housing and internal metallic components of his device were finished by the machine shop. However, he still had not obtained the specially shaped permanent magnets that were critical to the device's operation. The plans he purchased said that they could be purchased from a company that the original inventor had worked with when he built his prototype. When Bernard tried to contact the company he learned that they were no longer in business. Bernard then decided to call the inventor and tell him of the problems he was having in obtaining the special magnets. The inventor was surprised to hear the news and told Bernard that he would have to find another company to make the magnets for him.

Since he had already invested almost $1500, Bernard wasted no time calling every magnet supplier he could find in a desperate attempt to find the correct magnets for his copy of the free energy motor. His predicament was made worse by the fact that the magnets, aside from being composed of a very expensive rare earth alloy, had to have an odd and precisely shaped cross section. After another month of searching, he finally obtained magnets of the correct material and strength, but then had to return to his local machine shop for more help.

For a hefty price, the machine shop people said that they would try to grind his magnets until they were the correct shape he needed for his motor. This took another month, but the day finally arrived when Bernard's basement shop table was covered with all of the parts that the inventor had specified in the blue prints that Bernard had purchased from him almost six months earlier. Each part was a precise duplicate of the ones that went into the original motor and everything was ready for the final assembly phase of his project. However, the cost to Bernard had been quite high. In fact, the finally price was close to $5000! But, Bernard still considered it a bargain. After all, he was now one of the first persons to be part of the revolution in free power generation that was about to dawn for humanity.

The next night, before he began the final assembly, Bernard called me to tell me of the many and various problems that he had to overcome to finally reach the point where his dream of a lifetime was soon to be realized. We chatted for almost an hour long distance and the conversation ended with me offering him my admiration and a confession that I wished that I had been in a financial position to achieve what he was about to achieve. I also lamented that I could not be physically present with him so that I could personally witness the event.

Another week went by and one night my phone rang. It was Bernard again, only this time he sounded very depressed. He then told me that he had assembled the motor and checked to make sure all of its parts were in their precise positions. Then he manually grasped its output pulley and gave the lubricated and smoothly turning shaft a strong spin to get it started as had been recommended in the plans for the device. The rotor shaft spun flawlessly on its precision bearings and then, quite unexpectedly, gradually slowed to a stop! He tried again and again only to get the same result. It soon became obvious to him that the "motor" did *not* work! Bernard disassembled it and checked the positions and orientations of all of its expensive magnets, but everything was exactly as the plans called for.

The next day Bernard called the inventor of the device and told him of the results. The inventor expressed bewilderment over the outcome and suggested that the problem must be that the permanent magnets used were not quite right. Aside from this opinion, the inventor could help him no further.

I asked Bernard during a later phone conversation if the inventor had ever told him that he had a *working* model of the motor. Bernard said the matter had not come up because he had *assumed* that, with a patent for such a device, the inventor must have had one. At the time, I had also assumed that a working prototype existed and that it must have been produced and demonstrated in order to satisfy the requirements of the Patent Office.

However, after later research into this matter, we learned that it is quite possible to get a patent for what an *inventor* considers to be a perpetual motion machine *without* providing a working physical model to the Patent Office as long as he words the claims section of his invention's patent application properly. More specifically, one can literally patent *any* unique assemblage of weights, levers, springs, magnets, rotors, or circuits *without* providing a working physical model to the Patent Office just so long as none of the invention's claims stated in the application say or even suggest that it is actually *creating* energy during its operation.

Any invention that claims to do so is currently considered to be impossible by the Patent Office since that would be a blatant violation of the energy conservation law and such a claim becomes an immediate reason to reject granting the inventor a patent for his invention. To overcome that rejection and finally obtain the patent, the inventor must be able to demonstrate a working physical version of the invention to the Patent Office officials. To date, no one has been able to do that.

Whether or not the inventor of the "motor" described above ever had a working model of the device which he patented and whose plans he marketed, neither I nor Bernard will ever know. But this bitter experience taught both of us a valuable lesson which was that, ultimately, a granted patent really means nothing when it comes to determining the actual functionality of an invention that the *inventor* advertises as being a "free energy" or "overunity" or "perpetual motion" machine unless that detail is stated in the patented invention's claims and this claimed functionality has been sufficiently demonstrated to officials from the Patent Office which granted the patent.

Having learned from Bernard's sad experience, in the absence of a *granted* patent containing *specific* claims that it was actually outputting more energy than it consumed in its operation, I would only accept the reality of such a device if I had actually invented it myself or if another inventor of such a device let me personally examine, test, disassemble, and reassemble it to prove to myself that it was not a fraud. If this was not possible, then I might accept the evaluation of an independent committee of *trained* specialists if they verified that the device was actually genuine. I would also recommend to others that they likewise adopt such very strict standards when considering such devices, particularly if they are considering making any sort of financial investment in their development or promotion.

Bernard passed away several years ago without ever seeing the dawn of the energy revolution of which he so much wanted to be a part. The device he attempted to replicate has also faded into obscurity. The true believers in such inevitably disappearing inventions will claim that they were suppressed by devious governmental or commercial forces for a variety of reasons. The skeptics will simply point out that they disappeared for one simple reason: they did *not* work! I find myself in the skeptic's camp with regard to the many "revolutionary" devices that have come and gone over the centuries. They were quickly forgotten when it eventually became obvious to the people interested in them that their inventors were more adept at delivering excuses rather than working devices with some sort of commercial value.

With all this said, one might wonder where the author stands with regard to the possibility of constructing a working permanent magnet motor especially considering his early failed attempts to do so. Before I answer that, however, let me describe some of the actual physics of systems using magnets and then proceed to the implications of this physics.

Imagine that two magnets are positioned near each other so that their poles *attract* each other. If the two magnets are allowed to move toward each other, then they could be rigged up so that in the process they performed some work which is the same as saying that they *released* some energy to their environment. Perhaps they could have strings attached to them so that they both raise a small weight against the pull of Earth's gravity as they slide together and finally make contact. If they do not lift weights as they slide together, then they will both accelerate toward each other as their separate kinetic energies increase. Upon impact, they will become motionless, but their kinetic energies will not be lost. These energies will be turned into the sound energy emitted as contact takes place and thermal energies and the temperatures of both magnets will each increase, perhaps, by a fraction of a degree Fahrenheit.

Let us suppose that it is then decided to separate the two magnets and return them to their *original* starting positions. To do this, external forces will have to be applied to both magnets. Since forces are being applied to move the two magnets back to their original starting positions, work is being performed *on* the system which is the same as saying that the magnets are *extracting* some energy from their environment which could be the hands of the person pulling them apart again.

If careful measurements are made of the amount of energy that is released as the magnets come together and the amount of energy required to again part them and return them back to their starting positions, then it will be found that the two amounts of energy are *exactly* equal! In fact, the two amounts of energy would be equal *regardless* of the starting separation distances between the two magnets, the orientations of the two magnets with respect to each other before they are brought together or after they make contact, the strengths of the two magnets, or the materials that composed the two magnets. In fact, this situation persists regardless of how many magnets are involved. The amount of energy lost as a collection of magnets comes together will always be exactly equal to the amount of energy that must be supplied in order to restore them back to their original starting positions and orientations with respect to each other.

It is for this reason that the usual attempts to build a permanent magnet motor that uses conventional bar or disc shaped magnets are doomed to failure. In my early efforts to create such a device, I would use a rotor that carried one or more magnets and which would rotate inside of an outer stator which had various magnets fixed to it. I could usually get some rotation of the rotor, but it always reached some equilibrium position about which it would briefly oscillate before coming to rest.

I once tried a single fairly powerful magnet on a rotor that could turn smoothly through a 360° circular path inside of a stator that carried dozens of closely spaced magnets arranged to describe a mathematically perfect spiral about the rotor. Interestingly enough, this arrangement actually produced *no* motion in the rotor! It was when I saw this that I finally realized that my single rotor magnet was actually magnetically coupling with *all* of the magnets in the stator and not just the ones closest to it. I then had to accept the reality that a permanent magnet motor, if it was going to work, would have to have some very novel approach to dealing with the interactions taking place between its rotor and stator magnets.

After giving up on only using permanent magnets to construct their permanent magnet motors, many inventors think that they might be able to build such a device if they could impose a "shield" of some sort between some of the magnets in their invention when it comes time to pull the magnets apart so that they can return or "reset" to their starting positions. At first glance this seems like a workable idea. It promises to produce a device which releases energy as unshielded magnets are attracted toward and physically come closer to each other, but consumes less or, hopefully, no energy as the magnets are then pulled apart and return to their starting positions within the motor.

Many permanent magnet motors that use various types of shielding materials have been invented and, unfortunately, none of them worked. Most often when two magnets come together within such a device, a metal plate will suddenly be physically moved until it is interposed between the magnets in an attempt to cut off the attractive force acting between them. Designs using this principle range from devices that have swinging magnetic pendulums to rocking magnetic beams. I have even seen some devices that look like magnetic turbines wherein the shielding is wrapped around the trailing sides of the rotor magnets in an attempt to diminish a counter rotational attractive force acting on them and the rotor they are attached to as the rotor's magnets spin past the stator's magnets.

What makes the shielded magnet approach unworkable is the fact that the shielding material is just an easily magnetized sheet of metal alloy which, when brought near a magnet's pole, becomes magnetized itself and turned into a temporary magnet. While the shield may diminish the magnetic interaction between any two magnets in a device, this interaction is then merely replaced by two new interactions between the magnets and the shield. The end result is that there really is no reduction in the energy required to return the two magnets to some starting position after the shield is in position.

Realizing the limitations of magnetic shielding, a few inventors have taken the next step in reducing or eliminating the internal attractive forces that prevent their device's magnets from easily returning to their starting positions within the motor as its rotor spins. They simply insert electromagnetic coils inside the device which then, at critical intervals during the rotation of the rotor, provide electromagnetic fields that act so as to weaken or completely cancel out the attractive forces acting between any two separating permanent magnets within the device. This approach does work. Unfortunately, the motor they finally get running is not really a free energy device. It is merely an unusual *electric* motor!

The problem with such a hybrid device is that the electromagnetic coils it contains require a supply of electrical current in order to produce their electromagnetic fields. Obviously, this current could be produced by the device's rotor as it turns and then operates a small attached electrical generator. However, it seems that no one has ever built such a device wherein the energy required by its electromagnets was *less* than what the motor produced in operation so that there was some energy left over to perform useful work in the motor's environment. In fact, in the real world where energy is always wasted by friction and other inefficiencies, these devices always produce less energy then their internal electromagnets require. As a result, they quickly slow to a standstill after being given an initial electrical power "boost" to get them started and running at operating speed.

With all of the above analysis given for permanent magnet motors, the reader may believe that the author has reached the conclusion that such a free energy device is an impossibility and should not be pursued. However, this is not the case. I do still believe that a permanent magnet motor is possible, but that a design radically different from those usually tried by inventors is required.

In the last several years I have seen demonstrations of a few alleged permanent magnet motors that were very impressive. I recall one device, cylindrical in shape and about the size of a large microwave oven, which was demonstrated at a European engineering school to a group of students and professors there. The rotor shaft of this machine had fan blades attached to it and, when the rotor was released, it would spin with great speed. I estimated the power output in the hundreds of watts. To dispel suspicions of a hoax, after the demonstration, the inventor even dissembled the device and let the witnesses examine its parts! There were no hidden batteries or wiring and the device only contained hundreds of small rare earth alloy permanent magnets.

Possibly, the inventor, by carefully orientating the device's rotor magnets at a certain angle with respect to its stator magnets, found a design that managed to extract more than enough energy from the rotor and stator magnets as they parted from each other so that the rotor magnets could, as the rotor continued to turn, have more than enough angular momentum built up in order to overcome the repulsive forces resisting their return to their starting locations. As a result the rotor shaft would experience a nearly constant torque on it that would accelerate it to a high rotational speed. Whenever the fan attached to the rotor shaft spun, its air drag would then serve as aerodynamic braking so that the rotor shaft would not spin so fast that it might be damaged. Ordinarily this should not be possible, but, perhaps, with a special design it does, indeed, become a possibility.

In order to better understand the role of the proper angular orientations of the magnets within a genuine working permanent magnet motor, consider the following analogy.

Imagine a windmill with its flat panels attached to an axle that is supported by the central structure of the windmill. Imagine that on this special windmill the planes of the panels are adjustable so that one can angle them in different orientations relative to their axle.

If all of the panels are adjusted so that their planes are perpendicular to the axle that they are attached to and the incoming wind then strikes them at an angle perpendicular to their planes (which means the wind's direction of motion is aligned with the axle's length), then no torque will be created on the axle as the wind just smoothly flows around *all* edges of each panel and, consequently, the axle will not rotate. If, however, the panels are all slightly angled the same way with respect to the wind striking them, then suddenly the wind will tend to flow more around one edge of each panel (which becomes its trailing edge) than its opposite edge (which becomes its leading edge). This selective flow of the wind over the surface of each panel will then impart a force to each that will result in a torque being developed on the axle and the entire radial array of panels will begin to rotate as their axle rotates. This rotation can then be used to perform useful external work such as grinding various forms of grain into flour.

If this effect can be simulated using the repulsive forces acting between permanent magnets, then, perhaps, it might also be possible to make a permanent magnet motor's rotor shaft experience a constant torque that will cause it to spin at high speed.

Where would the energy such a device outputs come from? There is really only one possible source. As such a device ran and continuously outputted energy, that energy would have to be "paid for" by a slow but steady loss of the *mass* of the magnets in the device. According to the well known equation, $E = mc^2$, that Einstein popularized in the early twentieth century, all mass has a certain and usually very large amount of energy associated with it. Indeed, the two are really identical although the units used to measure them are different. One simple way to remember their interrelationship is to just state that wherever one has energy present he *always* has mass present and wherever he has mass present he *always* has energy present. If a permanent magnet motor is constructed that is undeniably outputting energy to its environment, then the mass of the motor's active moving components, its permanent magnets, *must* be diminishing over time. On the other hand, as that outputted energy powers some device external to a permanent magnet motor and causes that device's component parts to begin moving about, the energies and masses of all of those parts will be increased. These changes in mass, however, are usually not evident because, even for large changes in the energies of machine components, the associated changes in mass that are simultaneously taking place are very small and difficult to accurately measure.

While there probably are big changes soon coming in the field of free energy research involving permanent magnet motors, I want to use the remainder of this chapter to discuss another approach to building such a device which I have not seen previously attempted. I do this in an effort to give future permanent magnet motor builders a new direction to consider for their research efforts. In fact, this would be the direction that I personally would go in if I was still actively involved in attempting to build a genuine permanent magnet motor.

Let me begin by stating that I have noticed that all of the unworkable permanent magnet motor designs that I have seen in the past have one characteristic in common: they use *dipole* magnets. That is, their magnets all have both a "north" and a "south" pole. What, I've often wondered, would be the consequences of using magnets which only have *one* pole? Such a magnet would have either a north pole or a south pole, but not both and would be referred to as a "monopolar magnet" or just a "monopole".

Consider the following scenario. Imagine a large toroidal or donut shaped electromagnet with its wire turns wrapped around a hollow, air-filled toroidal core. This electromagnet is activated by allowing an electrical current to flow through its wire turns and, once this happens, a strong toroidal magnetic field is then established at its core. The magnetic field lines inside of the torus will be concentric with its core and each line will run completely around the core of this electromagnet to form a closed circle.

Now, let us assume that we can somehow introduce a magnetic monopole into the hollow core of the energized toroidal electromagnet. What would the result be?

Well, according to electromagnetic theory, the monopole would immediately feel a repulsive force on one side of it and an attractive force on the opposite side of it due to the electromagnetic field that fills the core of the toroidal electromagnet. These unbalanced forces would then cause the monopole to move rapidly around the hollow core of the torus and the monopole's motion would *never* cease! In other words, this arrangement would actually be a form of perpetual motion. If it was possible to somehow connect the moving monopole inside of the core to a generator outside of the torus, then this system would continuously produce free electrical energy. As I mentioned above, I believe that the only way to explain where this free energy came from would be to assume that in the process of generating it, the rest mass of the monopole magnet's mass possessing subatomic particles was slowly being extracted and then converted into the mechanical energy that drove the generator. As a result, over time the total mass and weight of the monopole magnet would decrease.

Quite unfortunately, however, magnetic monopoles are only theoretical abstractions and they do not exist in the real world. But, despite this I believe that it might be possible to approximate them with a permanent magnet. Imagine that we could take a small Alnico bar magnet (note that the acronym "Alnico" is derived from the magnetic material used to make these magnets which is an alloy of ALuminum, NIckel, and CObalt) and somehow make it as soft as putty. Now imagine that we were able to grasp the end poles of our putty magnet and then stretch it out until it was as long and thin as a strand of spaghetti. If we were then able to again restore it to its original alloy hardness and sharpen each of its pole ends, we would notice that *each* end of our stretched out strand-like Alnico magnet would have properties very close to that of a hypothetical monopole. Thus, its north pole end would be a point that *emitted* magnetic field lines in a nearly perfect radial pattern about itself and its south pole end would, similarly, be a point *into* which its magnetic lines of force could be envisioned to be entering or converging in a nearly perfect radial pattern.

Although we can not currently make alloys behave like putty (that may be possible when we are finally able to generate anti-mass field radiations having B rich character), we need to use another method to produce such a long, thin, straight, strand-like magnet. One simple method would be to obtain a long thin, but rigid, iron rod which could be about 6 inches in length and file each of its ends until they were pointed. While held with pliers, this rod would then be heated and, while it was still hot, placed between the poles of a large horseshoe or C shaped magnet. This will allow some of the magnetic field lines between the larger magnet's air gap to flow into and through the much smaller strand-like heated magnet that we've placed in the gap. (Note that if a horseshoe magnet with an air gap this size is unavailable, then one might substitute a magnetic "battery" which could be made from a stack of disc magnets 6 or more inches in height which has steel plates extending from both ends of the stack. These plates will then channel most of the stack's external magnetic field lines to flow between the two plates so that they behave as though they were being provided by the pole faces of a large horseshoe magnet.)

The hot strand-like magnet is then kept in position in the larger magnet's air gap and allowed to cool to room temperature. When it is finally at room temperature it can be removed and we will find that it is now powerfully magnetized. Its end that pointed to the horseshoe magnet's north pole will be a magnetic south pole and its other end that pointed to the horseshoe magnet's south pole will be a magnetic north pole. The ends of the newly created magnets that are north poles should be marked somehow so that they can be identified later and a total of eight of these newly created magnets should be made.

The magnetic field lines that leave our strand-like magnet's north pole and reenter its south pole will be approximately radially arrayed at both of the pointed ends of this newly magnetized thin rod of iron. Thus, each pointed end will behave like a single magnetic monopole and we will have two monopoles of opposite magnetic polarity at the opposite ends of our straight, strand-like magnet.

Now that we have something that approximates the properties of a magnetic monopole, we need to make something that will approximate the properties of a toroidal magnetic field. With these it should then be possible to construct a new kind of permanent magnet motor.

We can roughly simulate a toroidal electromagnet's core magnetic field that would exist in a short *section* of its torus by simply using a single bar magnet. As the magnetic field lines flow out of the bar magnet's north pole and loop around the length of the magnet back toward and into its south pole, they will all be heading in the same direction and will entirely envelope the bar magnet. If we were to place eight of these bar magnets around the circumference of a small circle and orient them all so that their north magnetic poles all pointed in the same *counter*clockwise direction, then there would be a region inside of the circle of magnets in which there would be some field lines that would form a weak circular magnetic field all of whose lines pointed in a clockwise direction.

This array of eight permanent bar magnets will serve as a "stator" and can be made using a piece of stiff cardboard. A circular hole about 16 inches in diameter is cut in the square piece of the cardboard and the bar magnets are then simply glued as near to the rim of the hole as possible so that the side lengths of these straight bars form tangents with the circumference of the hole. The magnets should be glued on one at a time and, aside from having all of their north poles pointing in the same counterclockwise direction, each magnet's center must be positioned so that, with respect to the center of the circular hole, it is located at 45 degree angular intervals away from the centers of its two adjacent magnets.

At this point we must introduce our approximate magnetic monopoles into this circular magnetic field which approximates a toroidal magnetic field. We will do this by constructing a simple "rotor" that will hold our eight north "monopoles" in a fixed configuration about an axle that is free to rotate.

To make our simple rotor, we can take a quarter inch diameter wooden dowel and cut it until it is six inches in length. We then hammer in two small steel nails into the centers of the two ends of the wooden dowel. Once the nails are firmly embedded in the dowel, their heads are cut off and the sections of the headless nails protruding from the ends of dowel are then rounded off with a file. These nails will later serve as metal pivots of our rotor. Next, we find a craft store that sells some sort of styrofoam plastic cylinder that is about 6 inches in diameter and a few inches thick. A hole whose diameter is slightly smaller than that of the dowel is drilled through the exact center of the styrofoam cylinder and the previously prepared wooden dowel is then forced through the hole until the cylinder is located at the center of the dowel.

Since the hole in the cylinder was slightly smaller than the diameter of the dowel, the cylinder should not slide about on the dowel.

Next, we take the eight previously prepared long, straight, strand-like magnets that we made and determine which of their ends are the north poles. Since these were previously marked, this is easily done. We then take the opposite *south* pole ends of each of these long thin magnets and insert them into the styrofoam cylinder that is pierced by the wooden dowel. It's very important that only the south pole ends of each long thin, strand-like magnet is inserted into the styrofoam cylinder and the eight magnets must all lie in the same plane such that each has an angular separation of 45 degrees from the two magnets nearest it. The magnets must be firmly embedded in the styrofoam and, if any are loose, some water based craft cement can be used to secure it.

We now have a simple rotor with eight north "monopoles" located at its outer circumference (the south monopoles are all clustered near the rotor's axle and are not used) and a simple stator that will approximate the toroidal magnetic field of a toroidal electromagnet.

The rotor will be a little over a foot in diameter and will be placed into the center of the stator's hole so that the plane of the rotor's strand-like magnets is in the same plane as that of the eight bar magnets of the stator. The center of the rotor axle must be located at the exact center of the stator's hole. The stator can be elevated several inches above a flat surface, such as provided by a sheet of wood, by gluing four pieces of carefully cut wood to the surface and then gluing the corners of the square piece of cardboard holding the eight stator bar magnets to the tops of the pieces of wood. The rotor must then be mounted so that its axle remains vertical inside of the hole of the stator and the rotor must be free to turn on its two rounded nail pivots. Something shaped like a tiny metal "cup" must be found (this cup might, perhaps, come from an old retractable ball point pen's metal top button) and then securely glued with its opening facing up to the flat surface which supports the square sheet of cardboard and its eight stator magnets so that the cup is located directly under the exact center of the circular hole in the square sheet of cardboard. The bottom nail pivot of the rotor's axle is then placed into this cup along with a bit of vaseline to act as a lubricant.

Finally, the rotor's axle must be carefully aligned so that it stands up vertically and some means must be found to provide support to its axle's top nail pivot. This can be done by constructing a rigid arm that will extend over the stator's hole until it reaches the rotor's top nail pivot. The arm need only have a small hole placed into it for the nail to be inserted. Perhaps a stiff piece of copper wire with a small loop formed into it to receive the rotor shaft's upper nail pivot would work. Again a bit of vaseline will serve as a lubricant. When this prototype "permanent monopole magnet motor" is finally completed, each of the rotor's eight north magnetic "monopoles" will, at its closest, be separated by about 2 inches from each of the stator's eight bar magnets.

What will happen when the rotor is placed into the stator and then released?

In theory (and I use the word "theory" because I have not personally attempted to construct this device), as soon as the rotor's north magnetic monopoles enter the circular magnetic field inside of the stator's circular hole, they will feel a small force acting on them that will cause the rotor to begin rotating in a clockwise direction. This is because the magnetic field lines emerging on the trailing side of each of the rounded tips of the rotor's strand-like magnets are immediately repelled by the stator's magnetic field behind that tip while those magnetic field lines emerging from the leading side of each rounded tip of the strand-like magnets are immediately attracted to stator's magnetic field lines ahead of that tip. The result should be a small torque which, if the rotor bearings are properly lubricated, will cause the rotor to begin rotating in a clockwise direction about its axle.

In order to make this device work, it is important to make sure that the rounded tips of the rotor's strand-like magnets, which serve as north magnetic monopoles, are not too close to the stator's bar magnets. If this is the case, then the rotor may only turn a small amount until the north monopoles at the tips of its strand-like magnets reach the south pole of each bar magnet of the stator. At that point the attraction of the monopoles for that end of each stator's bar magnet may become so strong that the rotor will become locked and unable to rotate.

The reader contemplating constructing this device should be prepared to move the bar magnets of the stator to various distances from the axle of the rotor in order to minimize any excessive attraction that is acting on the rotor magnets. If the strengths of the external circular stator magnetic field inside of the hole and those of the rotor's "monopoles" are just right, then, in theory, this design should work.

Unfortunately, I can not claim to be the inventor of the above described permanent monopole magnet motor. The idea for it occurred to me after reading about an ingenious inventor that may have been the first to use this concept in the construction of one of his working permanent magnetic motor devices. The details of the story are somewhat vague, but I will, as best I can recall, recount them here for the occasional researcher that may wish to delve further into this subject.

In the early 19th century there was an uneducated Scottish shoemaker named Spence (I think his first name was John) who, as a child, had developed a fascination with all types of magnets. When only an adolescent, he managed to build a rocking beam device which used permanent magnets as its sole source of power. After this, he brought out another model which used a swinging pendulum to operate. In both models, he had solved the problem of using shields to cut off resisting magnetic fields when the moving permanent magnets in the devices returned to their starting positions. He claimed to have a black substance which he had developed which could instantly block a magnetic field, yet was *not* itself magnetized in the process. If this substance was applied to the pole faces of a horseshoe magnet, then no external magnetic field could be measured outside of the magnet!

Slowly his fame as an ingenious inventor spread throughout Scotland and it was not long before he came to the attention of some professors at the famous Edinburgh University. They immediately decided, without even examining his devices, that they must all be fraudulent. However, they did invite him to come to the University to demonstrate his inventions so that they could see exactly what he had.

Several years passed before Spence was able to make the trip to the university. When he finally arrived, he is said to have done so while riding in a small horseless carriage that he had built. It was just large enough to hold two people and he offered to take all the professors who wished to try it for a ride. The important detail of this little story is that the carriage Spence drove the professors around the grounds of the school in was powered solely by a rotary permanent magnetic motor that the shoemaker had constructed. This motor had only *one* moving part which was its rotor. One by one the learned men of science took a ride in Spence's carriage and were totally bewildered as to how it was propelled. The physics they knew said that it was impossible, yet they could not doubt the evidence their own eyes. Spence at all times tried to answer their countless questions, but he was an uneducated country fellow who had developed his own language to describe magnetic effects. Unfortunately, his explanation of how the motive power source in his carriage operated was totally unintelligible to the professors.

Spence was later offered huge amounts of money for the secret to his devices and the formula for his black substance magnetic shield. However, he had no particular interest in wealth and seemed to more enjoy mystifying people with his marvelous devices.

Spence was a solitary person and had few friends. As another man in the prior century had been to gravity driven machines (i.e., Johann Ernst Elias Bessler), Spence, in his century, was probably the world's leading expert on magnetic science although all of his knowledge was of an instinctual nature gathered from much trial and error. Eventually, as he got older, he began to have health problems and was no longer physically able to construct his permanent magnet motors.

On his death bed, Spence entrusted a small box to one of his few friends and asked him to open it someday after he had died. Spence told the younger man that it contained something that he had built when he was a young person and just starting to experiment with magnets. The friend agreed to keep the box in his house, but eventually forgot where he had placed it.

Approximately twenty years after Spence's death, the friend found the small box that had been entrusted to him. He opened it to find that it contained a most unusual device. The box contained the brass balance wheel mounted in a pocket watch movement from which the hairspring, pallet lever, and remaining gears had been removed along with the dial and hands. On the balance wheel two small iron nails had been mounted so that their heads were near the staff of the balance wheel and their sharp tips were pointing directly away from the staff and out over the rim of the balance wheel. This balance wheel and the nails it carried could rotate freely through 360° of motion. Near these, a small horseshoe-shaped magnet had been fixed in place that had a small block of Spence's secret black shielding material inserted between its poles.

Upon opening the small box, Spence's friend noted that the balance wheel was rapidly spinning. When he stopped it with his finger, it would immediately begin spinning again when released. The friend then decided to dissemble the balance wheel assembly with the intention of cleaning and relubricating its two ruby jewel bearings. When this was done, he noticed that the jewels had been badly worn despite their prior lubrication. This indicated that the device must have been in continuous rotation for *decades*! When it was again reassembled after servicing, the balance wheel immediately resumed its rapid spinning motion.

Like many such devices which may have been genuine, Spence's early permanent magnet motor has been lost and never again duplicated. From its description, however, I am convinced that it was a simple, working monopole motor. Hopefully, the plans I gave above for a permanent magnetic monopole motor will, like this device Spence bequeathed to his friend, one day allow such devices to be built again.

Chapter 8

Artificial Intelligence and Consciousness

LIKE MOST "BABY BOOMERS", I was conditioned from an early age to believe that actual working robots would be invented during my lifetime. These creations were standard fare in the science fiction novels and movies of the '50's and '60's, yet it is now over a half century later and we still do not have anything that closely resembles what was promised back then in the crystal ball of science fiction.

While we do not presently have the "Robbie the Robot" from the *Forbidden Planet* movie or "Mr. Data" from the *Star Trek: The Next Generation* television show that would make this age old dream of a fully interactional mechanical man, it appears that the soon anticipated advances in computer technology and materials science may, indeed, eventually make that dream a reality. Even now, the budding science of "nano technology" suggests that it may soon (as in a decade or two) be possible to construct computers which will have the awesome storage capacity of the human brain without any of its biological frailties. It may even be possible to actually "grow" these artificial brains in various chemical solutions and then program them in seconds with a lifetime's worth of learning and a completely formed personality. Unlike a human brain, processes inside these artificial brains would take place nearly instantly.

Obviously, the next evolutionary step after the development of these artificial brains will be to find a way to house them in some sort of high tech android body. These can even now be fashioned from various silicone or organic polymer elastomers that are virtually indistinguishable by touch or appearance from human skin. Such an android body would have a central hydraulic pump that would send pressurized liquid silicone oil to hundreds of small slave cylinders located throughout its body. These slave cylinders would function like the muscles in a human body and allow for the smooth movement of the android's limbs. Hundreds of pressure sensors embedded in such an android's "skin" would instantly tell it how much pressure it was exerting on objects in its surroundings.

To power this first true android, it would seem that the soon to be developed fuel cells would be ideal. Thus, this machine would, like a human, breathe in oxygen containing air which would then be channeled into a battery of fuel cells along with a stream of hydrogen gas. As these two gases were reacted inside of the fuel cell battery, the electricity generated would power all of the android's internal equipment. The only "waste" created in this process is pure water. When the android was resting and only using minimal power, this water could be exhaled as vapor. However, after strenuous exertion, a large amount of water might be created and this would, when convenient, be emptied from an internal storage tank via a process that would mimic urination.

The structural details of such an android's body are, surprisingly, the least important features of such a device. The critical part that will make it all possible will, in fact, be the complexity of its artificial "brain".

Most of the devices available today which are referred to as robots are, in reality, nothing more than high tech puppets. They are basically remotely controlled devices which can only perform tasks in their environments so long as such tasks are carefully monitored and directed by a human controller. Thus, when one sees "robots" being used by the bomb squads to defuse a bomb or rovers moving across the surface of Mars to study its geological features, one should keep in mind that these devices are merely extensions of some remote human controller's body and mind. A "true" android would initially be given one or more tasks by a human being to perform in either oral or written form, but then all of the many complex sub-tasks involved in accomplishing those larger tasks would have to be performed completely and *independently* by the android. This is a critical condition that such a device must be able to meet before we could truly consider it to be more than just a mechanical puppet.

This now brings us to the subject of "artificial intelligence" about which there has been much debate during the last half a century. There have been several definitions of this including the famous one given by Alan Turning, a mathematician, philosopher, and early pioneer in the science of cybernetics. He once suggested that a machine could be considered to be intelligent if a human conversing with it over a teletype machine could not tell whether or not he was conversing with another human being or a machine.

Of course, we already have machines that have a very high degree of artificial intelligence. I am referring here to the familiar personal computer that almost every household in the developed nations of the world now possesses. Through its complex operating system program and additional programs that are temporarily downloaded into its "volatile" memory, such a device can execute complicated calculations or other data processing tasks in a fraction of a second. With the right programs, now referred to as "software", a computer can process images made up of millions of data points and detect patterns in them that would be imperceptible to the human eye. There are even programs now in existence that can "recognize" *spoken* words, analyze the meaning of those words, and give spoken responses that are appropriate under certain limited conditions.

However, all that has been achieved in computer science to date is still only about a fraction of a percent of what will be needed to build a truly independently functioning android. Computers will have to become much smaller, much faster, and contain programs of staggering complexity in order to do this. Only the field of nano technology would seem to offer any hope of achieving this goal.

Let us now consider what the advanced artificial intelligence of a truly independently functioning android might involve. This should give the reader an accurate idea of the complexity of the program its brain will have to contain and be able to use.

First and foremost, the android's brain must be simultaneously aware of the operational status of all of the many hundreds or even thousands of components that compose its body. Thus, it's brain must be able to sense such things as the temperatures of its internal components, the hydraulic pressures in each of the slave cylinder "muscles" which position and move its metal skeleton, the angular positions of all of its limbs, the electrical current output from its fuel cell battery, and the pressure of the gaseous hydrogen fuel stored inside of it's internal supply tank. The brain must also have a sense of balance so that, if the android starts to fall over, it can take immediate action to compensate for this and remain in whatever position it is supposed to retain. All of these various sensing and adjusting processes that will serve to maintain the android's internal homeostasis will, no doubt, require that a considerable portion of its artificial brain's programming be devoted to them.

Next, we must consider the android's senses which, as in humans, allow this device to perceive various objects and energy sources in its immediate environment.

The sense of touch can be easily created by embedding hundreds of tiny piezoelectric pressure sensors in the plastic skin of the android along with tiny temperature sensing resistors. Each such sensing unit will provide a sensation of touch to an area of skin of about 1 square centimeter and all of these sensors will have to be connected by microscopically fine and flexible wiring to the machine's brain. As in a human being, their function will be to tell the brain if too much and potentially damaging force is being applied as the android's limbs attempt to move objects in its environment. Temperature sensing is critical in helping the android avoid potentially thermally dangerous environments and, of course, contact with fire.

The senses of sight and hearing are, perhaps, the easiest to achieve because we already have a technology that can produce high resolution digital cameras and high fidelity microphones. Of course, the data streams coming from these miniaturized devices will have to be rapidly analyzed by the android's nano technology brain so that it will be instantly aware of what objects are located in its immediate environment. The programming for the android's eyes will be further complicated by the need to provide the android with stereoscopic vision. Not only must all objects in the images from each camera eye be identified by their shape, color, texture, etc., but their parallax shift with respect to background objects must be measured so that the brain can determine how far away the objects are from the android's body.

The senses of smell and taste will be the most difficult to equip an android with. Although our android will not need to eat food or drink beverages, it may occasionally pretend to do so when in social situations with humans and it having even rudimentary senses of smell and taste would allow it to make appropriate comments about food and beverages in social settings. This material would later be discharged, undigested, from a storage sack within its abdomen.

Eventually, however, it might be possible to construct some sort of digestive system that could extract the chemical energy from carbohydrates, fats, and proteins by reacting them with atmospheric oxygen inside of a "food fuel cell" inside of the android's body. This energy, in the form of electrical current could then be used directly to power the android or could be used to electrolyze some of the water its hydrogen / oxygen fuel cell battery normally produces back into gaseous hydrogen and oxygen again which would be stored and then used by the hydrogen / oxygen fuel cell battery at a later time to produce electrical power and liquid water again. The various waste products produced by the food fuel cell would have to be periodically removed from the android's body via defecation, exhalation, and urination in processes that would simulate what actual humans do. It seems that, again, nano technology will be able to eventually provide sensors that can detect various aromas and flavors. These will be super miniaturized versions of the military "sniffers" used to detect the presence of toxic chemicals in the air during chemical warfare.

As was noted above, when this android is given a task to perform by a human being, the robot must be able to break that task down into a finite series of sub tasks and perform each of these successfully in a logical and efficient sequence in order to finally complete its main task. The android must be able to follow either verbal or written instructions, be able to ask for additional information if needed, and be mobile enough to overcome various barriers in its environment that might hinder the performance of its current task.

Most importantly, the brain of a true android must be capable of learning or acquiring new information about its environment. Although it will initially be programmed with a large amount of data concerning objects on earth, the physics and chemistry of everyday situations, and the various rules of conduct in human society, a true android's brain will still have to be able to add to this database when it comes across objects and situations with which it is unfamiliar. These abilities will, most likely, represent the ultimate level in artificial intelligence that can be achieved by a machine.

The question quickly arises as to whether such an android would have consciousness as the term is generally understood by humans. The answer to this is rather complex. Obviously, the android's brain will be aware of objects in its environment and of the operational status of various parts of its internal mechanisms. However, in order to be conscious, it must also have "self-awareness": that is, the android must be aware of itself as *separate* from its environment. This ability would seem to naturally be a part of its complex programming because all distances to surrounding objects in its environment will be computed with respect to the position of the android.

Consciousness for humans also requires the ability to engage in imagination. Thus, we can engage in day dreaming in which we "call up" various mental images related to problems we are trying to solve. We then alter these images in different ways as we try to predict how the real objects that the images represent might behave in various situations. Often this process leads to novel solutions to the many problems of human existence.

We see from this that humans, whether awake or while sleeping and dreaming, can produce a kind of virtual reality representation of the external world within the hidden neural circuitry of their brains. Any robot who claims it is conscious would have to be able to basically do the same thing. Its brain would have to be able to generate a virtual reality model of its external environment within its nano circuitry and then manipulate the object images of that model to see if any novel solutions related to the performance of the android's current task were produced. In other words, our android would not truly be conscious unless it had the ability to exercise creative imagination.

Some readers of this chapter may feel that a machine could never achieve the kind of consciousness that humans possess. The final reality of the situation may be that androids will, eventually, achieve a kind of "super consciousness" that will actually *surpass* that of humans! Again, all of this becomes possible if the promises of nano technology can be kept.

A truly independently functioning android would be able to nearly instantly identify objects in its environment by measuring their various sizes, shapes, and color patterns and then comparing them to a catalog of stored object parameters within the android's brain. If the object was not in this catalog, the android would be able to access a worldwide data base using wireless satellite technology. This same technology would also let any two androids or group of androids communicate with each other via a special binary android language. Thus, a problem that might be beyond the analytical capabilities of a single android could be submitted to a group for analysis and recommended course of action. Humans already do something similar to this when they read and respond to posts on internet forum sites.

In attempting to perform a task, an android could literally contemplate thousands of possible ways to complete its assigned task. It would then only try to act on the methods which were the least expensive or least energy consumptive. To a human observer, such behavior on the part of an android would appear to be highly creative. Indeed, it seems logical to expect that androids would eventually be inventing most of the devices used by humanity. Androids could even be assigned the task of improving their own kind by designing more advanced nano technology brains that could hold an even more complex personality program. This personality program itself would eventually be created by android programmers working under the careful guidance of humans. Such a complex personality program might be customizable by the robot's owner so that the android would be able to get along better with its human owner. The owner of an android could take a personality test and the scores from its various sections would be used to suggest how to set the various personality parameters of the android brain's personality program so as to assure maximum rapport between human and machine.

There are predictions now that the first robots in the human home will be in the form of robot pets. Because a pet robot is smaller in size than a full human sized android, the cost will be less for the owner. Even a small robot pet might be able to perform many useful tasks around one's home. It should also be possible to construct robot pets that would have near human intelligence. This will take some getting used to on the part of humans, but robot pets capable of human speech that could hold an intelligent conversation with their owners would be an asset to lonely people who would normally keep a live pet for companionship. For older people, such a robot pet, perhaps in the form of a dog or cat, could monitor the owner and call for help in the event that the elderly person was incapacitated due to sickness or injury.

When the day finally arrives on which we have independently functioning androids walking the streets of our towns and cities as they go about performing their assigned tasks, some humans may be bothered by the status of these machines as servants in our society. However, I think one must always keep in mind that an android will only have the feelings or emotions that its human owner decides to allow it to have. Thus, it will not be angry that it must work 24 hours per day, 7 days per week. It will not feel that its rights are being violated or that it is being taken advantage of. It could only have such feeling if its human owner decided for some purpose to allow it to have those feelings. One could just as easily program such a machine to express happiness and gratitude for being assigned to continuous labor of the most menial sort. Such happiness would be absolutely genuine as far as the android was concerned.

The first truly independently functioning androids will, unfortunately, be very expensive. This cost will only be justifiable if they are required to perform tasks that are too difficult or dangerous for humans to perform. Hopefully, with mass production techniques, it will be possible to eventually lower the cost of an android to the point where the average person can own one. Such a device could provide companionship or perform various domestic services around the home which could include such things as cooking, cleaning, home repairs, lawn maintenance, laundry, etc. In short, all of the tedious chores that humans tend to avoid could be assigned to an android.

Although humans will acknowledge that androids are intelligent and even conscious soon after their appearance in our lives, I do not believe that these machines will ever be legally considered as another lifeform. They will therefore, ultimately, not be accorded any more rights than one's automobile or home computer. This status will not be offensive to the androids because, as was stated above, they can be programmed to willingly accept any role assigned to them in human society. In general, this role will be to serve the needs of their owners.

Of course, as with any potentially dangerous piece of equipment, it will be mandatory that strict safeguards be incorporated into them which will prevent them from being misused. Thus, no robot or android would be allowed to follow any instruction or continue with any task that its artificial brain's analysis indicated might lead to the injury or death of a human being or other higher lifeform. There would be strict penalties for any human being who attempted to bypass these safeguards.

However, one wonders how such an android would react to a physical attack upon its "person" by a human being. An android would be an expensive piece of equipment and it would seem logical to give it some ability to prevent itself from being damaged, destroyed, or stolen.

But, what if the *owner* of the android had decided for some reason to destroy the machine? Perhaps in such cases, the android would only take immediate evasive action to protect itself if attacked by any human other than its owner, but would submit to an attack by its actual owner.

What might an android do if its assigned task was to protect its owner from some other human being? In such cases, the android might only be allowed to use non-lethal force or weapons to perform its task. In these cases, the android would not be allowed to use clubs, firearms, or knives and would be restricted to the use of such things as chemical, electrical, or optical weapons such as those that produce a bright flashing laser light that can dazzle and stun an attacker temporarily.

The behavior of androids in such situations will pose a variety of moral and ethical problems that will have to be ironed out by the legal system and the companies that create the programs for these machines. Completely satisfactory solutions will not come over night, but may take many decades to achieve.

While one can easily imagine androids performing many of the tasks in a society that tend to be boring, dirty, or dangerous, their amazing intellectual abilities would also suit them well for performing many of the "routine" tasks that are now done by various professionals.

Thus, I can envision hospitals staffed by android doctors and nurses. A surgical team of androids would be able to perform the most delicate of surgeries in a matter of minutes and, thereby, greatly reduce the risk to the patient from the comatose state induced by anesthesia.

Android medical researchers might be put to work to develop some new lifesaving drug or procedure. Unlike humans, the androids would be able to work on the problem around the clock and day after day without the need to sleep or eat. They would only occasionally stop for a few minutes to replenish their internal supplies of hydrogen fuel. Once the company that owns the researcher androids had paid for them, that company would have a source of highly skilled labor that would virtually work for free. Of course there would be the occasional cost of maintenance for each android. But, not having to pay human researchers high wages, would help greatly lower the cost of new medications. This drastic reduction in the cost of development would help make these drugs widely available to even the poorest of countries.

It is easy to extend this scenario and predict that androids will eventually make human labor unnecessary. However, because of the high cost per android, I see little chance of this happening anytime soon. People will always want to make sure that somewhere in any organizational hierarchy that directs and controls android workers, there will be human beings in charge. Thus, no android or group of androids will ever be operating completely independently of human oversight. This will prevent the possibility of some situation coming into existence that might pose a threat to humanity. If such an unforeseen situation did ever occur, then it is important that human beings be able to override it and shut the androids down until the problem could be extensively analyzed and corrected.

So, when can we expect the arrival of the kind of android envisioned above?

Predictions in the fields of science are risky things at best to make because, more often than not, they tend to be wrong or way ahead of schedule. I remember that during the '60's the futurists were predicting that we would have lunar cities and regular travel to Mars by the end of the 20th century. Now that we are in the 21st century, these things are currently being predicted for the middle of our century. I am of the belief that, if we are restricted only to the use of rocket technology, even these latest forecasts are extremely optimistic and ahead of schedule (however, I do see them as possible for the middle of this century if the propulsion technology of UFOs is developed and made widely available for human use).

When it comes to android technology, however, I would be very surprised if the machines described above were not finally perfected by the middle of this century. The structural materials needed are already available, a miniaturized, high efficiency hydrogen fuel cell power plant is almost here, but, unfortunately, the nano technology needed for the android's brain is still only in its embryonic stage of development. This last element is an absolutely critical one that must be developed in order to make lifelike androids an everyday reality. Present day microprocessors simply do not have the data processing power needed to handle the enormous amounts of data that would be entering and leaving the artificial brain of an independently functioning, intelligent, and conscious android.

In the future, with the help of android workers, humans will be able to construct a world that is far safer, cleaner, and more comfortable and fulfilling for all of Earth's people than is our present one.

Chapter 9

Time Machines and Time Theory

Recently, I decided to ask some people if they could give me a definition of what time was and I received a variety of answers back that, in the final analysis, did not really define the word. Most people readily agreed that time was something that was measured with a clock, but, when it came to defining exactly what that "something" was, their opinions began to vary widely.

One fellow, who was familiar with Einstein's Theory of Relativity, said that time was the "fourth dimension" as though that pronouncement would somehow resolve the matter. There was also some talk about the past and the future and how time seemed to "flow". A few thought the present moment was like a car moving down a road: the stretch of road ahead was the future and the road behind the car was the past. One person pointed out that time was the amount of waiting one had to do between the occurrences of two separate events. As a future event "drew nearer" to the present, that person said that the time to it "grew shorter" and as a past event "receded" into the past, the time to it "grew longer". Thus, I learned that time was a measurable something that could vary in "length" depending on how it was measured. However, after my informal survey of opinions was completed, I could only conclude that the general concept of time is a rather muddled one which most people resolved by trying not to think too much about the subject! In this chapter, however, I want to explore the concept of time in depth so as to, hopefully, finally determine the true nature of this "something" which seems so mysterious and has perplexed philosophers and scientists throughout the ages.

Much of our difficulty in defining time stems, I believe, from the nature of the words we use to describe it and from the way in which the concept is presented in various literary works, particularly in science fiction stories.

Students of physics eventually learn that all physical units such as those that are used to measure things like force, motion, energy, temperature, pressure, field strength, etc. can ultimately be expressed in terms of only three fundamental "dimensions" which are: mass, length, and *time*. In fact, after one derives an equation in science that describes the relationship between the various parameters of a physical system, it is very important to perform what is called "dimensional analysis" on the equation to make sure that the various dimensions of mass, length, and time (and the exponents to which they are raised) that appear on one side of the equation are exactly the same as are those which appear on the other side of the equation. If, however, the dimensions which appear on each side of the equation are *not* identical to each other then either the equation is invalid or there is a math error somewhere in its derivation.

The dimensions of mass and length are easily understood by most people. Mass is a measure of the quantity of subatomic particles that are present in a piece of matter and length is a measure of the quantity of an arbitrarily defined span of space that can be fitted between two points in space. Time, however, is somewhat different from these dimensions because it involves measuring the quantity of *motion* of the universe that takes place between two distinct *events* in the universe in terms of an arbitrarily defined amount of motion of some readily observed part of the universe.

It is important to remember that when we talk about time and its measurement, we are always really talking about the measurement of motion of the universe. For example, when we say that one minute is made up of 60 seconds, we are really just saying that, during the unit of time we call a "minute", the universe undergoes 60 times as much motion as it does during the unit of time we call a "second". The unit of time we call a second was originally selected to approximately equal the amount of motion of the universe that takes place between two successive, readily observed events such as the swings of a clock's pendulum or the pulses of the human circulatory system. Currently, we arbitrarily define a second as 1 / 86,400[th] of the motion that takes place in the universe between two consecutive maximum angular displacements above the horizon of the Sun in Earth's sky as viewed from any point on the planet's surface.

Thus, we see that when measuring the time between two particular events that we are considering, regardless of whether these events are produced naturally or artificially, we are really determining how much motion of the universe has occurred between those events. Obviously, for ancient, pre-technological people on Earth this motion of the universe would have been one of the most obvious motions connected with the passage of Earth's days and months: the apparent rising and setting of the Sun and the various phases of the Moon. As technology was developed, these motions would eventually be replaced by the synchronized motions of the hands of mechanical clocks or by the changes in the displayed numerals of electronic digital clocks. Such artificial devices then indirectly allow the original source motions of the part of the universe that inspired them to be conveniently monitored even when the user does not have access to a clear view of the sky or when the astronomical objects involved are not visible.

How much motion of the universe will take place between two events is an important thing for a sentient living creature to know.

For example, if one knows that an anticipated event will not take place until several days worth of motion of the universe have occurred, then he would realize that he would need, perhaps, to eat a dozen times and have several sleep cycles *before* that future event occurred. If, on the other hand, the anticipated future event will not take place until only a few hours worth of motion of the universe has occurred, then one might be able to comfortably forgo both food and sleep before that event. In this example, the person is aware that his body's metabolic processes are just a subset of all of the motions of the universe and that if too much motion of the universe is required before an anticipated event occurs, then this also means that excessive motion in the atoms and molecules in his body that underlie his metabolism will also occur. This excessive extended motion of the atoms and molecules of the person's body could, if uncompensated for with sufficient rest and nourishment, lead to the discomfort of fatigue, hunger, and perhaps even death. Human beings have evolved so as to generally avoid discomfort and seek comfort and being able to accurately *predict* the amount of motion that will take place in the universe or the "time interval" between two events is an important part of avoiding both physical and psychological discomfort.

I have used the phrase "motion of the universe" repetitiously so far in order to try to emphasize to the reader that what we call time is nothing more than the quantity of motion and the resulting change in the positions of all of the component subatomic particles of the cosmos with respect to each other.

For example, when what we call an "hour" of motion of the universe has occurred, the Sun will have moved through about 15° of arc along its apparent trajectory across the our planet's sky as viewed from the perspective of a ground observer. However, it is not necessary for any visible macroscopic translational, rotational, or vibrational motions to take place in any part of the universe in order for time to be "flowing" or "passing".

Even if there were no such gross and obvious motions taking place anywhere in the universe, the very atoms and molecules present would, at the submicroscopic level, still be undergoing various types of motions such as molecular translations, rotations, and vibrations and would continue to do so as long as they did not cool, via the emission of electromagnetic radiation, down to the temperature of absolute zero which is about -459.69° F or -273.16° C.

However, even if any such measurable submicroscopic motions of every atom and molecule in the cosmos were to somehow come to a complete stop and the cosmos did, in fact, suddenly cool to a temperature of absolute zero degrees, time would still exist and be flowing because the various perpetual motions of the electrons and nucleons *within* individual atoms would continue without interruption. Indeed, and finally, should even the perpetual motions of all of the electrons and nucleons in the cosmos suddenly cease, then the existence of flowing time would still be supported by the *remaining* "intrinsic" motions of the "ultimate" component particles from which all electrons, nucleons, and other subatomic particles are themselves composed! We can think of these intrinsic motions as the ones experienced naturally by the smallest particles of matter that can exist. They are, therefore, the most fundamental motions of the cosmos. (Note that the "standard model" of current physics proposes that all subatomic particles are themselves composed of only a dozen smaller particles consisting of 6 "quarks" and 6 "leptons". I think the reality, however, is that even these component particles will eventually be seen to be composed of what I refer to as "ultimate particles" because these are, themselves, *not* composed of any smaller component particles.)

As the reader may realize by now, time and the various levels of motion of the universe which it represents are quite impossible to entirely eliminate. In order to eliminate *all* time *everywhere*, one would, literally, have to eliminate all of the infinite number of motions taking place throughout the cosmos right down to those intrinsic motions possessed by its ultimate component particles! Thus, there is really nothing that we can do to completely stop the flow of time throughout the cosmos.

Generally, there is also nothing that we can ordinarily do to affect the *rates* at which the various subatomic motions of the cosmos as a whole take place that define how fast time "flows". I am not referring here to the rates of motion of such things as chemical reactions or the speed of an electric motor or an internal combustion engine which we obviously can control, but, rather, to the rates of the intrinsic motions of the ultimate subatomic particles which compose the electrons and nucleons that then compose individual atoms in our cosmos. Indeed, it is these intrinsic motions which subtly determine the rates at which all "higher" level motions in the universe can take place.

The rates of these intrinsic motions of the cosmos' infinite number of ultimate particles are, for all practical purposes, beyond the ability of our current science and technology to affect even on the smallest and most local level. These motions are truly the clock at the heart of the cosmos and are currently tamperproof. This is probably a good thing because these most fundamental motions of the universe are the basic determinants of all of the known laws of physics and chemistry. Their integrity is the only thing that assures the eternal existence to the cosmos! The unalterable nature of these fundamental core motions means that during any particular second of time as much motion of the universe will occur as during any other second of time. One second of yesterday's time will involve as much motion of the universe, on the ultimate particle level, as one second of time measured tomorrow.

With these basic concepts of time theory in mind, we are now in a position to begin an analysis of the potential for the development of a so-called "time machine".

I recall going to see the movie *The Time Machine* when I was nine years old. It was the George Pal version of the novel written by H.

G. Wells and was released to theaters in 1960. Later, during the mid-1960's, I remember seeing it on our first color television. It is a powerful, dramatic story of a 19th century inventor who, out of sheer disgust for the warfare of his era, conceives and constructs a machine that allows him to escape into the "distant" future where he plays a critical role in saving the "good" remnant of humanity from a horrible existence. This movie and its 2002 CGI special effects remake are must viewing for all science fiction fans.

Both movies are based on the first novel written by H. G. Wells in 1895 when he was 28 years of age. At the time the novel was written, Wells was recovering from tuberculosis which, in those pre-antibiotic days, required months of isolation and bed rest. We can imagine the boredom he endured as he lay in bed day after day watching the hands of a nearby clock circle around its dial and, at night, viewing the slowly shifting stars of the night sky through his bedroom window. Perhaps he imagined how nice it would be if only he could have a control lever mounted on the side of his bed that he could push forward on and which would then make time *outside* of his bed race ahead so that his remaining months of confinement could be completed in a matter of minutes! This scenario, with a fabulous time machine substituting for a bed, is depicted at the beginning of the novel and, very dramatically so, in both movies.

Wells' novel is probably, along with the works of Jules Verne, one of the most important in not only science fiction, but general literature as well. The reason for this is that the time travel concepts, language, and scenarios it presents have much to do with shaping the currently popular notions of the nature of time. To finally decide just how much of the "Wellsian" approach to time travel is plausible, let us now briefly analyze just how a real time machine would have to operate.

Just as our analysis begins, however, we immediately encounter a major obstacle to the operation of such a time machine. In order to "travel" into the future, such a machine would have to be able to do one of two things. Either the machine would have to be able to force the rates of the intrinsic motions of *all* of the ultimate particles that compose the subatomic particles in the rest of the universe *outside* of the device to dramatically increase or the machine would have to be able to dramatically decrease the rates of intrinsic motion of the ultimate particles that compose the subatomic particles of its operator's body. I must immediately dismiss the first mode of operation of a hypothetical time machine as being physically impossible because of the reasoning given previously, but what about the *second* mode of operation? Surprisingly, this second mode of operation does offer some faint hope for constructing a genuine working time machine.

However, when considering any time travel theory based upon the adjustment of the rates of intrinsic motion of ultimate particles, in no case would it be possible to construct a machine that would allow its operator to travel *backwards* in time so as to be able to journey into the past. This impossibility of backward time travel immediately eliminates the annoyingly illogical "time paradoxes" that such travel would give rise to and, of course, invalidates any science fiction stories that use this concept (which is probably over 90% of such stories!). This prohibition of backward time travel is also in accord with one of the consequences of Einstein's Theory of Relativity which implies that, regardless of the relativistic conditions involved, no cause can ever precede its effect. Thus, causality must always be preserved throughout the cosmos and one should never be able to construct any type of device which would be able to travel back to a state of the universe that existed *before* the device was activated to make such a trip.

Therefore, in the remainder of this chapter, we shall develop the concept of a time machine that would only be able to dramatically slow the rates of the intrinsic motions of all of the ultimate particles that compose the subatomic particles of the various atoms contained in the body of the device's operator when he is inside of the machine and has no effect upon the rates of intrinsic motions of ultimate particles composing objects outside of the machine's influence.

Exactly how this might be done is something I have occasionally considered, but about which I have reached no firm conclusions. Indeed, it might truly be impossible in which case this chapter is purely academic. However, as we learn more about the ultimate particles that compose the subatomic particles from which all matter is built, some way to affect the rates of their intrinsic motions may eventually be found. Most likely, as in the case of negating the mass of a hovering UFO, it will involve the generation of some sort of new and previously unsuspected field effect. Perhaps, as we begin to generate anti-mass field radiation routinely on our planet so that we can construct our own UFO-like air and spacecraft, we will find some way to use it to create this additional field effect and that will then allow the machine described below to be constructed. With luck, such a genuine functional time machine might be possible before the end of this our 21st century.

Let us now imagine that the operator or time traveler must climb into a chamber inside of our hypothetical time machine, close its hatch, and then activate and control it with various buttons and a large lever placed in front of his seat. The machine also has several windows built into it through which the operator can view the external world surrounding him.

If such a machine is possible and it is stationary with respect to the Earth's surface, we can now further proceed to imagine what might happen if it was activated and the operator moved its control lever forward just a little so as to then reduce the rates of the intrinsic motion of the ultimate particles of his body and the other interior parts of the machine to $1/100^{th}$ of their normal values.

This would then result in all of the biochemical and physiological processes in the operator's body slowing down to $1/100^{th}$ of their normal rates because all "higher" particle motions on the atomic and molecular level in his body are, as previously mentioned, dependent upon the rates of the intrinsic motions of its many, many component ultimate particles. Since all of the mental processes of the operator including his senses would now be slowed to $1/100^{th}$ of their normal rates, he would then perceive all processes he viewed taking place *outside* of his time machine through its various windows to be occurring 100 times *faster* than they were taking place as viewed by an outside observer.

For example, the hour hand on a nearby clock outside of his time machine would appear to the operator to move from one hour marker to the next in only 36 seconds and he would confirm this by observing his watch's second hand which would show that only 36 seconds of time had passed inside of his time machine. Like his body and his mind, his watch's second hand would actually be moving at $1/100^{th}$ of its normal rate of motion. To him, however, neither his bodily motions, thought processes, or his watch's second hand would appear to be moving at $1/100^{th}$ of their normal rates, but, rather, at their usual normal rates. Thus, he always sees all motions in the world outside of his time machine moving at 100 times their normal rates while all motions inside of his time machine seem completely normal to him. In reality, however, it is only the time traveler's motions that have been altered.

By pressing his control lever forward a little more, the operator of this hypothetical time machine might further reduce the rates of the intrinsic motions of the ultimate particles composing all of the subatomic particles of the atoms of his body. When these rates reached $1/1000^{th}$ of their normal values, the operator would perceive each 24 hour day outside of his machine as passing in only 1.44 minutes according to his watch whose rate of timekeeping would, again, be the same as that of his greatly slowed physiological processes. At all times, as he reduces the intrinsic rates of motion of all of his body's ultimate component particles, he will perceive his own bodily processes, movements, senses, and thoughts to be completely normal and only the motions of the outside world to be greatly accelerated when, in reality, it is only his processes and any inside of his machine which are not occurring at their normal rates.

As outside observers viewed our time traveler through his machine's windows, however, they would see an unusual sight. They would see him continuously, but he would appear to be moving at only $1/1000^{th}$ of his normal rate of motion. For all practical purposes, he would appear frozen in place and only by taking photographs of him every several minutes and then comparing them would outsiders even realize that he might actually be moving about inside of his machine.

In order for the operator to listen to what people outside of his active machine were trying to say to him, he would have to record their speech and then play it back at $1/1000^{th}$ of the rate at which it was recorded otherwise he would only hear a high frequency squealing sound. On the other hand, in order for the machine's operator to speak to people outside of his active device, they would have to record what he said and then have its rate of playback increased by a factor of 1000 or it would be inaudible to them. Obviously, this method of communication would be difficult at best, but it would be much easier for the time travel to understand what outsiders were trying to tell him than it would be for the outsiders to understand what he was trying to tell them because of the delay that they would always experience as they attempted to record and then accelerate even the shortest of messages from him.

If the machine's operator was to suddenly (to him, that is!) thrust his arm and hand through an open hatch or window and thereby placed it outside of the active region of the device in which the intrinsic motions of its component ultimate particles were suppressed, then that part of the operator's body would instantly regain is normal rate of physiological function. As a result, it would, in less than a minute of *his* time, be a biologically dead limb! The reason for this is that, once outside the machine, the limb would then only be receiving $1/1000^{th}$ of the normal rate of blood flow needed to maintain the life of its various cells which, if they were kept there for only ten seconds of his time, would be starved for oxygen for 1000 seconds or 2.78 *hours* of outside time! Just suddenly withdrawing his limb back into the machine so that the circulatory needs of its cells were reestablished would not restore them to normal again and, eventually, he would have to have the limb amputated.

The above described hypothetical time machine would also need to be isolated from the surrounding environment during the entire time that its operator used it to "travel" into the future. Ideally, since it is stationary, it might be placed in a natural cavern where it would be protected from the weather and disturbance by curious people.

If the time machine uses electrical power to produce the effect that greatly slows down the intrinsic motions of the ultimate particles which compose the subatomic particles of the operator's body and the interior portions of the machine, then a reliable, long-lived power source must be available to it and the best option would seem to be the use of nuclear batteries. It is also important to note that while the intrinsic ultimate particle motions of the operator's bodily atoms and the interior parts of the chamber in which he sits would be greatly reduced during the operation of the machine, large parts of the machine would *not* be so affected. In fact, such a time machine might not function if all of its atoms ultimate particle intrinsic motions were greatly slowed down.

For example, the time machine will have to have some sort of externally located clock that measures the passage of real time in the outside world so that a control computer will know when it's time to shut down the machine's intrinsic motion dampening field after the operator inside of the machine had been affected by it for, perhaps, a "trip" of several outside centuries into the future. Such a clock would, most likely, have a highly accurate quartz movement that would be used to control the duration of the long journey to within an accuracy of a few minutes of outside time.

Obviously, all of the materials used in the construction of the "outer" portions of this hypothetical time machine, which are unaffected by its ability to slow down all intrinsic ultimate particle motions inside of itself, would have to consist of materials that could resist corrosion and any deterioration in their structural strength as the centuries passed at their normal rate outside of the device. This suggests the use of thick ceramic materials and metal parts that have been heavily plated with noble metals such as gold or platinum.

One major advantage of this hypothetical time machine is that it does have a sort of built in, automatic "failsafe" feature. Should its operator set its controls so that he can complete a journey of, say, one thousand years into the future and, while this trip was in progress, there was then some sort of malfunction that shut down the power to the equipment that inhibits the normal intrinsic motions of the ultimate particles composing his bodily atoms as well as those of the interior of the chamber in which he is located, then all of these particles would instantly regain their normal rates of motion. Once that happened, he could then safely exit his time machine and see if he could diagnose the cause of the problem and correct it.

In the event that the quartz movement timer on the outside of the time machine somehow malfunctioned so that it was unable to reach a preset shutdown time for the machine, then the operator might have to use an emergency shutdown switch located *inside* of the chamber that contained his body. If that switch should fail to open, then he might be tempted to just open the hatch on the chamber and try to jump free of the machine.

Such a maneuver would, however, be a fatal one because, as was noted above, if a part of his body hangs outside of the time machine, then it will receive a blood flow that is far less than that of the remainder of his body which is still inside of the machine. Should the operator's head emerge from the active time machine first, he would quickly experience brain death because the bulk of his body still inside the machine's active operator chamber would contain a heart that was only beating at $1/1000^{th}$ of its normal rate and, thus, would only be able to supply his brain with $1/1000^{th}$ of the blood flow needed to keep its cells alive as they emerged from the machine's interior and immediately resumed their normal rates of metabolism.

Obviously, to eliminate this danger and allow for a safe emergency shutdown of the time machine by its operator, several switches could be wired in series. If one should fail to open, then there would be several others any of which, when opened manually, would immediately shut down the power for the time machine and thereby prevent the possibility of "runaway" time travel.

Unlike the time traveler in H. G. Wells' classic work, our hypothetical traveler would only be able to make *one* way journeys into the *future*. While he would, of course, enjoy the excitement of exploring time as he periodically emerged from his machine and left its remote cavern to, perhaps, visit a local community to see how society and technology were evolving, the knowledge he gained from such a visit in *his* present time would have no value for the people of the time which he had previously left. At best, any knowledge he gained during any of his stopovers would only be of interest to future historians or archaeologists. In each era that he stopped off in, he would be a person out of place (and time!), a sort of temporal drifter, who might attract undesirable attention from local inhabitants that feared his presence or even tried to interfere with the continuance of his journey.

Before concluding this chapter, I want to briefly mention another method of time travel that emerges from Einstein's Special Theory of Relativity. Although this method is quite impractical for realistic time travel, it is often cited as a solution that is merely awaiting the necessary technology to make it available. Let us first, however, very briefly summarize some of the postulates and consequences of Einstein's great discovery in the realm of physics.

Albert Einstein published his "Special Theory of Relativity" at the beginning of the 20th century and it was a breakthrough in rationalizing certain paradoxical results that had been obtained in various physics experiments conducted in the late 19th century. Basically, his theory starts with only *two* postulates (which are statements accepted as true although they are not themselves derivable from any "higher" theory). The first postulate is that the velocity of light (which is about 2.99×10^{10} centimeters per second or 186,000 miles per second in a vacuum!) has a *constant* value regardless of the motion of an observer *relative* to the source of the light. The second postulate is that all of the known laws of physics must remain the same in a particular "frame of reference" regardless of whether they are determined by an outside observer who is stationary with respect to that frame of reference or are determined by an outside observer who is moving with respect to that frame of reference. (Note that "frame of reference" here simply means an imaginary three dimensional coordinate system containing the object under observation which is used to determine its three dimensional shape, but can also include an imaginary stopwatch that measures the time interval between events involving the object and some means of determining the mass of the object.)

These postulates seemed obvious to Einstein and, as was briefly mentioned above, seemed absolutely necessary in order to neatly explain the strange and unexpected results that were obtained in earlier scientific experiments that had been conducted by other physicists.

Once he had accepted these two basic postulates as true, Einstein was able, through a variety of ingenious "thought experiments", to show what the consequences of these postulates would be. At the beginning of this chapter it was noted that all of the various physical units of measurement (such as for force, velocity, acceleration, etc.) can be rewritten in terms of the fundamental physical "dimensions" of only mass, length, and time. Einstein wanted to see what the effects of his two postulates would be on these basic physical quantities and what emerged is considered to be one of the most important theories in all of science.

His theory describes exactly what happens to the fundamental properties of mass, length, and *time* as an object with these properties has a force continuously applied to it in order to accelerate it long enough so that its velocity begins to approach that of light. The results are rather bizarre and paradoxical and this fact accounts for the great resistance to the Special Theory of Relativity that initially followed its publication. However, during the last century, many independent physics experiments have been conducted concerning the Special Theory of Relativity and, so far, every one of them has verified the conclusions Einstein reached as to what will happen to the mass, length, and time of an object whose velocity approaches that of light at which point the object is said to have a "relativistic velocity". It took almost a quarter of a century after its publication before the world of science finally acknowledged the validity of Einstein's research.

Initially, as energy is supplied to an object in order to accelerate it, the velocity of the object will increase at the *square* power of the energy supplied to it. Thus, when the velocity of the object is low, doubling the amount of energy supplied to it will increase its velocity by a factor of four while tripling the amount of energy supplied to it will increase its velocity by a factor of nine.

However, one of the strange conclusions of the Special Theory of Relativity is that as the object begins to accelerate to velocities that are ever closer to that of light, less and less of the energy supplied to the object will go into increasing its velocity and more and more of it will go into increasing its *mass*! The equation that describes this effect indicates that in order to achieve the velocity of light, a material object (that is, one which had some mass to start with when it was at rest) would need to be supplied with an infinite quantity of energy and would then, once it had finally reached the velocity of light, possess infinite mass! Since this is clearly an impossible situation, the theory seems to prove that no material object can ever attain the velocity of light or exceed this natural universal velocity limit. Indeed, to date, no one as ever found or been able to create a material particle that moves at or exceeds the velocity of light. (There is, however, one very important exception to this scenario which, as I describe in my trilogy devoted to UFO technology, applies only to material objects, such as UFOs, which have been artificially rendered massless through the use of anti-mass field generators. These craft, while in such a condition, can then easily be accelerated to velocities far in excess of the velocity of light.)

The fundamental dimension of length is also altered dramatically for objects moving at relativistic velocities or velocities close to that of light. In these cases, the Special Theory of Relativity states that their lengths will actually physically contract along the direction of motion of a moving object. For example, if a spacecraft was to fly past the Earth at, perhaps, 0.8 times the velocity of light, then earthly astronomers observing the spacecraft's near approach would see it greatly compressed along its trajectory of motion. If such a spacecraft was normally cigar-shaped when landed and, thus, stationary with respect to the astronomers, then it might appear spherical in shape as it flew past the Earth at relativistic velocities. If the vehicle could achieve 0.99 times the velocity of light, then the astronomers might see it as an almost flat disc moving past the Earth! The plane of this disc would be perpendicular to the direction of its motion.

If the earthbound astronomers had a telescope with sufficient magnification and resolving power, then they might actually be able to see inside the spacecraft through a window or porthole in its hull. If they could, then all objects within the craft, including the crew, would also appear flattened in their common direction of motion. Interestingly enough, this length contraction effect is mutual in nature; that is, as the spaceship flies past the Earth, *its* crew will see the Earth moving at the same speed the craft has only in the opposite direction. This will then make the Earth appear to the crew members to be contracted along its direction of apparent motion. If the spacecraft travels past the Earth at close to light velocity, then our planet might appear to its crew to be a thin disc with only two circular sides!

Both the earthly astronomers and the astronauts in this thought experiment will, at all times during the flyby, perceive themselves as having their normal dimensions and claim it is the other's length that has been contracted along its direction of motion. Paradoxically, *both* the astronomers on Earth and astronauts in their moving spacecraft are *correct* in the claims they make even though these claims contradict each other! This mutual distortion effect is not just some sort of optical illusion, but is a real *physical* change in shape that is being measured by each set of observers in the other's moving frame of reference!

If one of the astronauts dons a spacesuit and crawls outside onto the hull of his speeding rocket ship with a tape measure and then proceeds to measure its length from its rear edge to the point of its nose (I'm assuming the craft has a typical three or four finned, single stage '50's sci-fi movie look to it!), then he might note that its length was 150 feet which is exactly what it was when the vehicle was resting back on Earth prior to its takeoff. He would say that the astronomers who are somehow measuring its length through their ground based high power telescope must be wrong when they radio the ship and say that their telescopic measurement of its length is only, say, 75 feet in length. But, what the astronaut does not realize is that, because of his high velocity relative to the Earth his ship is passing, he is measuring his spaceship's length with a tape measure whose actual stretched tape is now only 50% as long as it originally was back on Earth!

The astronaut has no way of knowing that he, his ship, and *everything* in it are now actually 50% shorter in whichever of their dimensions points in the direction of the craft's motion. As far as he can tell, his tape measure looks perfectly normal to him, but he is viewing it through eyes that have also been dimensionally shortened by 50% in the direction that the craft and his eyes move in and this unnoticed detail is what is making the tape's stretched out length look normal to him even though he receives a radio message from the earthbound astronomers telling him that the 150 feet of measuring tape he stretched out between his spacecraft's rear end and its pointed nose's tip only appears to be 75 feet in length to them.

The astronauts may also make observations of the Earth's surface as their spacecraft flies past our planet. If they have a telescope powerful enough, they may train it on the astronomers' observatory and then radio down to them that, using the equipment aboard their rocket to make their measurements, the circular base of the observatory's cupola or domed top, which was also exactly 150 feet in diameter when they were back on Earth, now appears to be flattened into an oval whose shortest diameter is only 75 feet across. Both the astronomers and astronauts claim that it is the other's structures that have become distorted. So, one might wonder who is actually correct.

The answer to this riddle is that, while both groups are making accurate measurements, it is really only the astronauts who are experiencing the relativistic change in their dimensions along the length of their direction of travel because it is only they who are actually moving near the velocity of light and not the Earth or the rest of the cosmos. When light from the astronomers' observatory reaches the spaceship, that light must then enter the volume of space containing the moving ship and its telescope which has become highly distorted due to their motions. It is that spatial distortion which makes the Earth and its ground based observatory appear distorted to the astronauts. But, this distortion is not just a mere optical illusion because there is no possible experiment that the astronauts could ever perform *while moving* at their near light velocity relative to the Earth that would prove that the observatory's dome actually had a circular base. Only if both groups have knowledge of Einstein's Special Theory of Relativity and awareness of their relative motion with respect to each other will they realize what is really occurring and it would then be the astronauts who would concede that they are the party whose lengths had become shortened by 50% from their normal stationary ones as far as sentient beings in the still more or less stationary cosmos outside of their ship were concerned.

As soon as the spacecraft lands back on Earth and it and its crew become stationary with respect to the astronomers, any previously mutually measured contractions in lengths of their structures along their apparent paths of motion will immediately disappear and each group will again view the other's structures as having their normal proportions. At this time, both the astronauts and the astronomers will agree that the rocket ship is 150 feet in length and that the observatory's cupola has a circular base that is also 150 feet in diameter. We see from this that once any relativistic motion between two objects ceases, the predictions of relativity theory no longer apply to them.

The most bizarre consequence of the Special Theory of Relativity is what it predicts must happen to the *time* durations necessary for various physical processes to occur in two different reference frames that are moving, especially at relativistic velocities, with respect to each other so that observers in *each* reference frame will simultaneously observe the same physical laws being obeyed in *both* systems. In this case, the theory predicts that the apparent time for physical processes to take place will actually "dilate" or expand so that an outside observer of a reference frame he has under observation will see processes taking place in it seem to take longer and longer to occur as the relative velocity of that observed reference frame with respect to his reference frame approaches the velocity of light. It will appear to the outside observer that the flow of time in the observed moving reference frame is slowing down and, if it was possible for that observed reference frame to reach the velocity of light (which is not possible as was stated earlier) relative to the observer's reference frame, then all time and all motion in the observed reference frame would appear to come to a complete halt.

Again, as in the case of length contraction, this effect of time dilation is *not* an illusion, but a real *externally* observable physical effect! And again, this effect is always *mutual*. Observers in each of two different reference frames moving at relativistic velocity with respect to each other will claim that the "flow" of time in their particular reference frame is quite normal but that it is all of the physical processes occurring in the other reference frame that are experiencing time dilation. And again, the observers located in two such reference frames in motion with respect to each other will, paradoxically, *both* be right about the claims they make!

Once again, let us consider the now famous thought experiment in which a spaceship leaves Earth and begins traveling around our solar system at near light velocity. As it travels about, Earth based astronomers again use their powerful telescopes to follow its progress and note that time aboard the craft is extremely dilated or seems to be flowing very slowly. They also note that the astronauts inside of the vehicle appear *almost* motionless. Occasionally, when the spaceship flies nearer the Earth, the astronauts use their own onboard telescope to observe that life on a flattened Earth also seems to be moving in slow motion and perhaps they calculate that, at its much slowed rate of rotation, it would take a week of ship time for the Earth to complete a single rotation! Their tour of our solar system is continued for several weeks of the *ship's* time at which point it is decided by the crew that they will return to Earth and land their ship.

Upon arriving back on Earth, the spaceship crew will have a surprise waiting for them. While the crew has only aged a few weeks as indicated by their ship's clocks and calendar, the people they left behind on Earth may have aged anywhere from a few years to a few decades depending on how close to the speed of light the ship was able to travel and Earth's various clocks and calendars all will also be years to decades *ahead* of those of the spaceship!

The crew of the spacecraft has, effectively, traveled forward in time in terms of Earth's time because the astronauts have not aged at the same rate as the astronomers they left behind when they first began their tour of the solar system. While it is true that both the astronauts and the astronomers *both* saw the other's rate of time flow slow down, the relative motion of the spacecraft at near light velocity with respect to the Earth and the rest of the more or less stationary cosmos forces the greatly slowed rate of flow of *ship* time to prevail when both the spacecraft and the Earth finally come to rest with respect to each again after the ship lands. Once again, this time dilation effect is real and has been observed in many physics experiments involving various electrically charged subatomic particles that have been artificially accelerated to relativistic velocities using machines known as "accelerators". The results of these experiments in particle physics are further verification of the reality of the paradoxical predictions of the Special Theory of Relativity.

In the above thought experiment, our spaceship was obviously functioning as a time machine and could allow its crew to move ahead in Earth's time by years to decades depending on the ship's velocity for each week of shipboard time that passed. The reader unfamiliar with the many problems of spaceflight may believe that the time dilation effect predicted by relativity theory provides a practical solution to engaging in time travel. Unfortunately, it does not. The best space technology we now possess will only allow us to accelerate a massive object to a velocity of about 50,000 miles per hour which is equivalent to about 13.89 miles per second. Since the velocity of light is about 186,000 miles per second, this means that our best current technology would only get our spaceship time machine up to a velocity that was 1/13,391th of the velocity of light which is less that 0.01% of the velocity of light! For all practical purposes, our fastest spaceship is virtually standing still compared to the motion of light and any time dilations that would be produced aboard it as it travels at its maximum velocity would be almost negligible as far as allowing its crew to "journey" into the future.

At this point, the reader may suggest that all that would be required to remedy this velocity problem would be to equip the spaceship with some sort of exotic nuclear or ion propulsion system that could operate continuously at low thrust so as to, eventually, accelerate the ship until it finally achieved the relativistic velocities needed to make forward time travel possible. Again, there is a problem that makes this solution unusable.

There are reliable calculations which indicate that a conventional spacecraft would not be able to exceed a velocity of about 20,000 kilometers per second or about 12,500 miles per second. At about this velocity the spacecraft would be impacting so much interstellar gas and dust that *all* of the thrust of its engines would have to be used to counter the drag on the ship's hull caused by this material and no further acceleration would be possible. Additionally, at this velocity, a collision with a single micrometeor the size of a pea would most likely destroy the entire spacecraft!

At a maximum velocity of only 12,500 miles per second, our hypothetical spaceship time machine would only be moving at about 1/15th or 6.7% of the velocity of light. The equations of Special Relativity Theory indicate that such a very low relativistic velocity would only dilate time aboard the spacecraft by about 0.23 % which is virtually negligible. For every 168 hour week spent aboard a spacecraft limited to moving at about 12,500 miles per second, the crew would find that they only "traveled" about 23 *minutes* into the future each time they landed the craft back on Earth. If all of the astronauts left Earth at the age of 18 and each managed to live to be 100 years old before they decided to land their spacecraft back on Earth, then they would find themselves to have traveled forward in time by about only 68 days or a little over two months time. It would hardly seem worth being cooped up inside a small spaceship for 82 years for so short a trip into the future.

From the above, the reader can appreciate why I am not an advocate of the various relativity approaches to time travel. They are not now and, most likely, never will be a practical solution to engaging in forward time travel and there are no scenarios in which they would permit backward time travel.

While there can be no doubt that the Special Theory of Relativity is one of the most important contributions to the structure of modern science because of its revelation about the time dilation effect noted by *outside* observers of reference frames moving at relativistic velocities, the theory fails to ask and answer an even more fundamental question about this strange effect: why should the motion of a reference frame cause time to slow down in that reference frame? Just stating that this effect *must* exist so that the laws of physics can be described with the same equations for two reference frames in uniform motion with respect to each other is really not a satisfying answer.

It is my belief that this effect of time dilation is somehow connected with the same effect that causes the mass of objects to increase as they are accelerated toward the velocity of light. Perhaps as kinetic energy is imparted to an object to accelerate it and this energy begins to show up as an ever increasing mass of the object as its velocity approaches that of light, that increase in mass somehow causes the rates of intrinsic motion of all of the ultimate particles which compose the object to begin slowing down. This effect then is relayed throughout all of the "higher" motions of the subatomic particles, atoms, and molecules that compose the object.

If the object is a living organism such as a human being, then his entire metabolism will slow down including the neurological processes responsible for his cognition and thought. Since all parts of a person's body experiencing such time or, more accurately described, *motion* dilation are equally affected, that person is completely unaware that all of his bodily processes have slowed down. He feels completely normal, but becomes aware of his situation only after he again becomes motionless with respect to the rest of the cosmos and can then measure the physical differences, primarily aging, that have accumulated between objects in his reference frame and those that are in the rest of the cosmos. I believe that these differences are simply due to the relativity induced differences in the rates of intrinsic motion of the various ultimate particles that compose the objects in his moving reference frame, including his body, and the rates of intrinsic motion of the various ultimate particles that compose the rest of the more or less stationary cosmos.

In order to make a kind of "unidirectional" Wellsian type time machine possible and practical, we need a way of producing the same kind of time (or motion) dilation effect predicted by the Special Theory of Relativity for moving objects *without* the necessity of accelerating the body of a time traveler to near light velocities. Possibly, at some time in the future, a chance discovery in a physics laboratory will provide a clue as to how to artificially create localized time dilation using some combination of electric, magnetic, or electromagnetic fields. Thus, as I previously suggested, it will be a new type of field effect that acts directly upon the intrinsic motions of the ultimate particles from which all subatomic particles are composed. But, should that discovery ever be made, the time traveler would still have to cope with all of the problems discussed for such devices in the earlier part of this article.

There are also a variety of other modes proposed for forward only time travel that occasionally surface in the literature of the subject which I will just collectively lump together as "suspended animation modes". They involve either completely stopping a person's metabolism by cooling his recently deceased body to temperatures near absolute zero (-459.69°F) using liquid nitrogen vapors or just greatly slowing down a still living person's metabolism by putting him into a drug induced coma while using ice packs to lower his core body temperature to tens of degrees below its normal temperature (about 98.6°F). Once in such a state, the person would either not age or do so at a much slower rate so that when he was eventually revived and regained consciousness, he would perceive himself to have jumped forward into the future.

About the first suspended animation mode of forward time travel mentioned above, I can only note that much experimentation in cryogenics research involving both dead and living organisms has been conducted since the early 1960's and, so far, no one has been successful in reanimating or reviving a "higher" organism that had been cooled to such low temperatures. I doubt if the process will ever be made workable because, even when attempts are made to "perfuse" the organism with a substance such as ethylene glycol in an attempt to saturate the bodily cells with an antifreeze that will prevent damaging intercellular ice crystal growth from occurring as body fluids freeze, the fact is that the antifreeze does not reach all of the structures within a cell uniformly. Upon freezing, some ice crystals do still form and critically damage delicate structures and membranes within the cells. When later attempts are made to thaw out such an organism, the then warm remains tend to form a kind of "soup" made up of damaged cells incapable of living again.

The second mode of suspended animation forward time travel does give one a better chance of being revived than cryogenic suspension at liquid nitrogen vapor temperatures, but it is still a very risky procedure that requires continuous monitoring by medical personnel. As with any kind of an induced coma, the risk of permanent damage to the body and brain increases with the length of time spent in such a state. While this technique is currently useful in certain types of surgery, I do not see it as feasible for a solitary time traveler wanting to journey decades or centuries into the future.

Finally, I shall conclude this chapter with a summary of what I have come to believe is the true nature of time.

I eventually came to realize that our perception of time is rather delusional in nature. There really are no such things as past and future states of the cosmos that somehow *now* coexist with the present state and to which we can instantly travel with a time machine. We basically live only in a present state of the cosmos whose component particles are constantly changing positions with respect to each other according to the various laws of physics.

The basic physical processes of the cosmos take place with rates that are ultimately determined by the rates of the intrinsic motions of the ultimate particles from which all of the higher particles of the entire cosmos are constructed (these "higher" particles being all of the subatomic particles, the atoms they form, and the molecules these atoms form). Thus, when we measure the "time" that takes place between two events in the cosmos, we are, usually without even realizing it, actually measuring the quantity of intrinsic motion that occurred in the individual ultimate particles of the cosmos during the "temporal interval" between the two events. Since we can not directly view and measure the intrinsic motions of these smallest possible particles, we substitute some other physical process for them whose motion we can easily observe and whose motion is consistently synchronized with the motions of these ultimate particles; that is, we use a reliable timepiece such as a clock or watch.

Since we will never be able to control the *directions* and rates of the intrinsic motions of all of the ultimate particles that compose our infinite cosmos and thereby make all of the time in the outside cosmos "flow" backwards (that is, by reversing their natural forward directions of motion and accelerating them) while simultaneously preserving the directions and rates of the intrinsic motions of our own bodily ultimate particles (and thereby allowing our bodies' time to "flow" forward at its usually rate), it is, therefore, logically impossible to build a machine that would allow *only* us to travel backward in time. However, it may someday be possible to create a unidirectional Wellsian type time machine that would allow its operator to greatly suppress the motions of the ultimate particles that compose the atoms of his body so that he would not appreciably age while the rest of the cosmos outside of the chamber containing his body continued to age in its usual direction and with its usual rates. That direction and those rates are determined by the still unaffected intrinsic motions of all of the cosmos' component ultimate particles since they are outside the influence of the time traveler's time machine.

Should working time machines ever become possible, the operator of such a device would need to seriously consider whether or not he would be able to make and then keep the commitment to engage in what would be *irreversible*, one way travel to future states of the cosmos and what value such exploration would have to the people he might meet along the way. His adventures and the knowledge gained from them would be of no use to the people he left behind, but might be of some benefit to future historians or archaeologists.

Finally, it should be remembered that, although a temporal explorer may be able to overcome the prison of time that limits those who do not possess his technology, he would still be subject to the same aging processes that they experience. Should some future time traveler spend twenty years of his life exploring the future, he will be twenty years older when he ceases his exploration. Unfortunately, the ability to travel through time, interesting as it might be, does not bestow immortality upon a time traveler and, in time, he will, unless he has some technical means of rejuvenating himself, eventually suffer the same fate as all of those he has left behind in his distant past.

Chapter 10

Jesus and the New Age

I HAD INITIALLY DECIDED NOT TO include a chapter dealing with religion in this volume. This was because I thought some might think it inappropriate for a book with the word "science" in its title. However, as anyone who has read both the Old and New Testaments of the Judeo-Christian Bible will have noted, that book does contain many descriptions of paranormal phenomena and so, since I am a student of such events, I decided its inclusion herein could be justified.

This chapter, however, is divided into two sections that take very opposite positions on the nature of the Jesus story. The first section, which might be considered a "pro-Christian belief" one, was originally written as an article for a website in October of 2004. In it I attempted to support the reality of this ancient story in terms of various New Age concepts, especially those that involve paranormal phenomena. The second section, however, is decidedly dubious of the story as it is generally accepted by the faithful and, although written at the present time, was the result of certain evidence I became aware of in and after the year 2007.

I present both views only to show how my personal opinions of the matter have changed during a time period of approximately one decade. It is ultimately up to each reader to decide for him or herself which, if either, of these polar opposite views is the closest to the truth of what actually may have happened about two millennia ago on the other side of our planet.

Let us now begin with my pro-Christian belief section.

About a decade or so ago, I remember reading of a study that was done to determine who the most influential people were in history. The study produced a list of the names of the ten individuals who had affected the greatest number of human lives on our planet. While I do not remember all of the names on the list, I do remember the top three individuals. They were Jesus Christ, Sir Isaac Newton, and Mohammed. Jesus' name was at the top of the list.

At the time I found this rather interesting because, aside from the New Testament, we know very little about Jesus and there are only a handful of historical references to him outside of the Bible, the authenticity of which are still being debated by scholars. Thus, we have no precise images of Jesus, no writings directly attributed to him, and only fragmentary knowledge of his life and teachings. Yet his story, recorded in the synoptic Gospels, has endured for about two millennia now and is a source of inspiration and hope for over a billion of Earth's current inhabitants.

In this, the beginning of the third millennium since the birth of Jesus, there seems to be a worldwide decline occurring in many of the traditional religions, Christianity included, which is unprecedented in history. Actually, the decline started somewhere back in the 15th century as the Renaissance or "rebirth" in classical learning and science got underway. Now, an ever growing percentage of people in the Western world profess that they do not really have any "strong" religious beliefs and some claim outright that they have no religious convictions at all and that God is not real to them.

While this movement away from the formal, traditional religions has increased the freedom of individual thought and action, it has, quite unfortunately, resulted in a general decline in the moral and ethical values of people and the societies they form. Indeed, since the Renaissance period, there has been a steady increase in the frequency of wars and the death tolls associated with them. In the last century alone the various "-isms" designed to finally perfect mankind have probably accounted for the deaths of over one hundred *million* innocent men, woman, and children. In these cases, the moral restraints required to prevent these disasters were tragically deficient.

In consideration of this slow moral decline, I have always wondered if it might not be possible to create a kind of composite philosophy or religion that would preserve many of the values of the formal religions, yet would not conflict with the many scientific revelations that have emerged since the Renaissance. This belief system would not be some strict set of guidelines, but, rather, would be a general philosophy of life and of the world which could serve as a guiding light for humanity as it enters the third millennium since the time of Jesus.

Slowly, over the years, I came to the opinion that certain aspects of the so-called "New Age" movement have the potential to fulfill this role. New Age philosophy seems to replace formal religious dogma and ritual with a gentler and more tolerant "spirituality". Although this term is hard to precisely define, it does imply certain concepts. It suggests that all human beings are brothers and sisters who, despite differences in skin color and language, share and are jointly responsible for the upkeep of the same speck of dust planet in an infinite and eternal cosmos. It also embraces the concept of "karma" which finds its origins in the Eastern religions and philosophies that actually predate both Judaism and Christianity.

A belief in Karma postulates that every thought, word, and deed of a person has potential consequences for either good or evil. It further suggests that if a person does good acts, he can expect good things to come back to him in greater measure and if a person does evil acts, then he can expect evil things to come back to him in greater measure. Thus, with the concept of karma, the person does not actually do good acts to appease one or more gods so he can gain the benefits of their favor and the good fortune that might bring, but, rather, because the cosmos is so structured that such behavior is in the individual's and the world's best interest. It is almost as though karma was the manifestation of some sort of cosmic "pleasure principle" based on morality and ethics. A sensible person will, therefore, do good acts because humans have a natural genetically programmed neurological tendency to seek pleasure and avoid pain.

I have also found that the spirituality of New Age philosophy is, in general, very compatible with the teachings of Jesus as found in the Gospels. In fact, in my study of Christianity, I soon realized that practically all of the events of Jesus' life and his teachings can be neatly rationalized in terms of New Age concepts. So, in the remainder of this chapter, I will briefly cover my own experiences with Christianity as well as the life of Jesus and his teachings and how these can be interpreted so as to fit in with the now emerging philosophy of the New Age.

I can begin by stating that I was baptized into the Catholic faith shortly after birth, but stopped attending Sunday masses at the age of eleven. I did not really miss weekly church attendance at the time because the masses were still being conducted in Latin and were unintelligible to me. As a result, I never received Confirmation as did the other Catholic children in my working class neighborhood. While they attended our local parish's Catholic grammar school, I attended the public grammar school down the street. To this day, I still feel uncomfortable in churches and will only attend Christmas and Easter services when pressured to do so by friends.

As a youth my only knowledge of Christianity came from the various religious movies that were shown during the holidays. Films like "King of Kings", "Quo Vadis", and "The Robe". From these I learned that Jesus was a real person who could perform miraculous feats of healing and promised that if someone had faith in him and believed he was the son of God, then that person could somehow also achieve immortality. The exact details of how this immortality would be achieved were, however, somewhat blurry in all of these movies. In later life I realized that this might have been done intentionally by the screenwriters of the movies. Such films are subject to considerable criticism by various Christian denominations so that, in order to make a film that will appeal to as many Christians as possible, a conscious effort was made to make Jesus' teachings as general as possible.

During the mid-1980's various support groups were in vogue and I found myself attending several of them regularly. Some were ministries operated by Catholic parishes in nearby towns and I found that, despite my lack of regular church attendance, I was made to feel welcome at them. Often priests or sisters would address our groups and at those times matters of faith would sometimes be discussed.

One Christmas holiday season I made up my mind to actually read my way through the New Testament so that I might, finally, get an accurate description of Jesus' life and his various teachings. I had been required to read several books from the Old Testament in a college course titled "Introduction to English Literature" and found the Middle English of the King James version of the Bible rather cumbersome. So, for my new study of the "holy writ", I managed to obtain a version of the New Testament called "The Good News Bible" which has the text written in contemporary English. Over the course of the next several

Christmas holidays, I managed to reread the entire New Testament several times over.

Finally, I reached a point where I was able to discuss the life and teachings of Jesus authoritatively with members of the clergy. It was then that I started to realize that much of Catholic dogma does not exactly agree with the Gospels! Being somewhat of a purist when it comes to such matters, I found this a little disconcerting. But, apparently, what Catholicism teaches about Jesus and the Bible is a mixture of the Gospels themselves, various oral traditions, and later clerical revelations concerning their meaning. From this, I began to realize just why there had been so many "schisms" within European Catholicism over the last two millennia and why we now have so many different Christian denominations!

From my many readings of the New Testament, I finally got what I think was an accurate picture of Jesus and his teachings. For the reader who is unfamiliar with Christianity, I will now give a short description of what I interpreted the "good news" to be.

Basically, somewhere around the year 4000 B.C. God created the universe, our Earth, its life, and the first two human beings, Adam and Eve. They were supposed to be immortal gardeners who would tend to the plants and animals in a paradise-like region of Earth called the "Garden of Eden". One day Adam and Eve were tempted by an evil fallen angel (in the form of a talking serpent) called Lucifer to disobey God. Their ensuing act of disobedience so angered God that he then punished Adam and Eve by casting them out of Eden. To make matters worse, God revoked their immortality so that they would have to struggle for the rest of their limited lives to grow food if they wanted to keep from starving to death. They could only continue the human race by having sex and producing children who, like their parents, would also have limited lifespans and be forced to struggle for their existence.

Four thousand years later (or two thousand years before the present time), God decided that mankind had enough of this harsh treatment. God would forgive the human race for the original sin of disobedience by Adam and Eve. However, this forgiveness was conditional. It would require a *human* sacrifice and a very special one at that. In fact, the person being sacrificed would be a *mortal* son of God himself! Those humans then living who believed that this person was, in reality, the son of God would, themselves, be allowed to achieve immortality. Of course, it was not quite so simple. Those who believed that the person sacrificed was the son of God also had to demonstrate this by following his moral and ethical teachings for the remainder of their lives.

Sometime during the summer of the year 7 B.C. a teenage peasant Jewish girl whose name (in English) was Mary was selected by God to bear the male child who would, as an adult, become the human sacrifice required for the redemption of mankind. At the time she lived in the Judean town of Nazareth and was betrothed to an older man whose name (again, in English) was Joseph. She was visited by angels one night and, as a result, became pregnant. Since she was a virgin at the time, this caused much distress for her family and future husband. But, the husband had a dream wherein God told him that he should tolerate the situation and Mary and Joseph finally married and looked forward to raising their first child who the Christian world would know by the name of Jesus. Most likely, Jesus was born around mid April of the year 6 B.C. One of the Gospels indicates that Mary eventually produced six more children with Joseph as the father. Thus, Jesus eventually had four younger half brothers named John, Joseph, Judas, and Simon and younger two half sisters.

Jesus grew up and helped his adoptive father Joseph in the family business which was either carpentry or some sort of general construction work. The next time we read of Jesus, he is about 12 years old and obviously well informed on the religion of Judaism. In the city of Jerusalem he disputed various religious issues with the rabbis at the Temple which was the center of Jewish religion and politics at the time. At a local wedding, he performed his first miracle by converting ordinary water into wine for the guests. He is a robust, intelligent young man who is apparently aware that he has a special mission and role to perform on Earth for God.

At this point in the synoptic Gospels, there is an approximate 18 year gap in the story of Jesus! When he is next described, he is a man over the age of thirty years old. After a 40 day period of fasting and wandering about in the desert, Jesus was baptized by his cousin and boyhood friend, John the Baptist, in the river Jordan. After this ritual cleansing, witnesses claimed that a dove descended out of the sky and landed on Jesus while a voice from a cloud announced that Jesus was the Son of God!

After this Jesus began an itinerant ministry of teaching and healing as he wandered about the countryside surrounding Jerusalem. He was joined by twelve men, referred to as the "apostles", who were convinced that, because of the various miracles he was performing, he had to be the "messiah" that was predicted to come in the Old Testament and who would help restore Judea to the dominant status it previously enjoyed under its former kings like David and Solomon. At the time, Judea was occupied and controlled as a vassal state in the Roman Empire. This was offensive to the Jews of the day who wondered why God would allow something like that to happen because it was a clear violation of a "covenant" or promise that God had made to the prophet Abraham about 2000 years earlier concerning the future prosperity and well being of the Jews in Judea (which is now called the state of Israel). The apostles believed that Jesus would use his apparently miraculous powers to somehow get the Romans out of Judea.

Jesus continued his ministry for several years and visited many small towns throughout Judea. He performed many miracles of a paranormal nature that include such things as healing, resurrecting the dead, levitation, mind reading, etc. which he said were are not really being performed by him, but rather by God! After he had gained people's attention with his miraculous acts, he would then deliver his message to them which he claimed was given to him by God although he never elaborated on the precise details of how he received this message.

Basically, the message stated that those who had faith that he, Jesus, was, in fact, the son of God and who followed his moral and ethical code of behavior would be granted *physical* immortality on Earth! He further stated that he would be killed and, as a final proof that he was who he claimed to be, would be resurrected from the dead by God! He would then depart the scene for a short and unspecified time period after which he would return in a glorious "Second Coming". During that visible worldwide event that would occur *during* the lives of some of Jesus' contemporaries, Satan and his demons, who tempted people to do evil things that were offensive to God, would be defeated in a tremendous battle of Armageddon. Once Satan and his demons were finally defeated and safely locked away in a deep pit, the dead who had believed in Jesus would then be resurrected and granted physical immortality. Those who were still alive when Jesus made his triumphant return would also be given immortality.

Jesus would then become the king of a worldwide kingdom and his throne would be located at the Temple in Jerusalem. This kingdom would, literally, be a paradise on Earth where all human needs would be fulfilled. Nobody would be sick, hungry, or in need and the whole globe would finally be at peace! However, after a thousand years of this blissful state, Satan would be freed from his pit and again be allowed to influence people. This would be one final test to see just who deserved to be part of the paradise that Earth would remain forever. Those who failed the test would be cast into a lake of fire, known as "Gehenna", on a terrible day of final judgment. However, they would not experience eternal punishment in some perpetual hell, but would just be burned up and thus annihilated for all time to come.

The survivors of this final judgment would then be granted a very special privilege. God, himself, would descend to Earth in a giant flying city, referred to as the "New Jerusalem", which would settle on the Earth. Then for the rest of eternity, Jesus, God, all of the good angels, and the surviving immortal humans would live happily forever!

Unfortunately, this message (also known as the "Good News") that I have described above is not delivered as quickly and distinctly in the New Testament. A reader must be prepared to study the *entire* text and piece together the various verses that describe the message. No wonder it tended to get muddled in the various religious movies that I enjoyed as a youth.

As Jesus continued with his earthly ministry, he eventually came into conflict with the religious authorities that operated out of the Temple in Jerusalem. Jesus considered these men to be a bunch of hypocrites who were more interested in their social status then in loving their neighbors and doing good deeds as Jesus claimed God wanted them to do. The religious authorities, who consisted mainly of a group of seventy rabbis called the "Sanhedrin", considered Jesus to be a troublemaker who was starting to inspire the Jews to take up arms and revolt against their Roman occupiers. Since these high Jewish religious authorities were responsible to the Romans for maintaining order, if Jesus' growing popularity resulted in an uprising, then there was the real possibility that the members of the Sanhedrin might wind up being tried and executed by the Romans!

Just prior to the Jewish Passover one year, Jesus, while in the company of three of his apostles, had a most interesting "meeting" known as the "Mount of Transfiguration" incident (see Matthew 17:1). Jesus left them, climbed a nearby hill, and suddenly his face began shining brightly as the hill was mysteriously enshrouded in mist! His apostles, who remained at the bottom of the hill, claimed that two strangers suddenly appeared out of the mist and they could see Jesus conversing with them. When Jesus returned to the apostles they asked him if he had just had a meeting with Moses and Elijah who were two important prophets from the Old Testament. They weren't, but were angels or messengers from God that had just told Jesus that he was soon to be executed in Jerusalem and that, after three days of lying in his tomb, Jesus would be restored to physical life to prove the authenticity of his message to the people of Judea. Naturally, the apostles were stunned by this revelation.

Jesus then entered Jerusalem and celebrated Passover with his apostles. One of the apostles, Judas Iscariot who handled the donations the group collected, left the dinner and eventually led the Temple guards to Jesus who, after the dinner, was waiting for them in an olive grove just outside the city walls. Jesus was captured that night and brought before various religious and political leaders. At the insistence of the Sanhedrin members to King Herod and then the Roman military governor of Judea at the time, Pontius Pilate, Jesus was accused of blasphemy, heresy, and treason and condemned to death by crucifixion.

Many attempts have been made to calculate the exact date that Jesus died on the cross. The fact that the Gospels say that a total solar eclipse occurred on that date which was also a Friday has allowed only two possible dates to be determined. They are April 7th, 30 A.D. and April 3rd, 33 A.D. If the first date is accurate, then Jesus would have been 35 years old when he died. If the second date is accurate, then he would have been 38 years old at the time of his death.

After being on the cross for several hours, Jesus finally succumbed to the loss of blood and respiratory distress and died. A Roman soldier involved in the execution then drove a spear head into Jesus' chest to make sure he was dead. The body was removed from the cross and taken to a nearby borrowed tomb that had been carved into the side of a rocky hill where it was anointed with various spices and then wrapped in linen. The tomb was then sealed by a heavy circular stone which was rolled into place over the entrance and guards were posted outside of it to make sure that the apostles could not steal the body and then claim that the predicted miraculous resurrection had taken place.

With the death of their leader, the apostles became very depressed and spent the next several days in hiding in Jerusalem. Two days later, some of the women who were associated with the small group decided to visit the tomb. They arrived at dawn on Sunday morning to find the tomb open and empty! The women then hurried back to report what they had seen to the apostles. This news produced great surprise and joy in the apostles and their faith in Jesus was reinvigorated.

For about 40 days after his resurrection, Jesus appeared several times to the apostles and other people. Finally, Jesus met with the apostles one last time, charged them to promote his teachings to the world, and concluded his interaction with them by levitating up into the sky! He then disappeared from their view as he merged with a glowing cloud that then floated away.

With the above, the reader has the basic story of the New Testament. The various epistles or letters to Jesus' early disciples that follow the four synoptic Gospels of Matthew, Mark, Luke, and John go on to describe the trials and tribulations of the apostles as they tried to promote the story and teachings of Christianity throughout the first century Mediterranean world.

It was after my first few readings of the New Testament that I realized that there was a *very* obvious problem with it that most Christians completely overlook. This problem involves Jesus' major prophecy during his ministry which was that his Second Coming was imminent and would be occurring sometime *before* the last of the original twelve apostles had died off (consider, for example, Jesus' statement in Matthew, Chapter 24, Verse 34 in the King James version). Jesus' Second Coming was to be a dramatic, worldwide event that would begin a paradise-like existence *on earth* for his devoted followers. The importance of this prophecy is attested to by the fact that there are over 1500 references to it, either direct or indirect, in the books of the New Testament. That works out to one reference every 25 verses on average! So, there can be no doubt that this prophecy is central to Christian belief.

The problem is that it has now been about two millennia since the times of Jesus and there has been no Second Coming! In other words, the major prophecy by the central figure of Christianity has proven to be a *false* prophecy! Such an obvious failing would probably tempt most non-believers to dismiss the entirety of Christianity (and, perhaps, any religion) as just being a hoax with no basis in reality. I, however, am not convinced that this is the case at all. A simple dismissal of religious beliefs as having no basis in reality is an approach which I think is untenable for a variety of reasons.

Let me start by saying that I have no doubt that Jesus was an actual living person and that he was relaying messages from a "higher power" to mankind as best he could. However, due to the nature of the communications that Jesus received, it is quite possible that he misinterpreted what he was told and it is this which is the reason for the eventual blatant failure of his major prophecy. It is also possible, of course, that there was no misinterpretation on the part of Jesus, but that somehow the higher power he was in contact with changed or even cancelled its plans with regard to humanity and this accounts for the failed major prophecy. My feeling, however, is that the "misinterpretation hypothesis" is the more probable one.

In the remainder of this first section of this chapter, I shall attempt to give a reinterpretation of the Jesus story told in the Gospels in light of the misinterpretation hypothesis. Along the way various New Age philosophical concepts will be used in the reinterpretation and, hopefully, the reader will begin to see some of the many parallels between early Christianity and the now developing New Age movement. I believe that this reinterpretation will not in any way diminish the important role of Jesus in history, despite his failed prophecy. Quite possibly, this approach may actually reinvigorate those whose faith has waned in our present secular times.

We can begin with the virgin birth of Jesus. In the Gospel account we find that in order for Jesus to be a perfect sacrifice to redeem all of humanity for its sins, he must, himself, be devoid of the "Original Sin" he would have acquired if he had been naturally conceived by a human father. This sin was one that originated with Adam and Eve when they disobeyed God and was automatically passed down through all of their descendants. It was believed that it was this sin that was the reason people naturally aged and eventually died. Without it, a person would actually be immortal unless, perhaps, his body was completely destroyed in some accident. To eliminate this sin in Jesus, God carefully selected an adolescent Jewish girl to be his mother because she was, herself, free of this sin due to her unique piety and purity. Thus, the girl, Mary, was visited by an angel who told her that she has been selected to bear the son of God. She was so overwhelmed and honored by this proposal that she immediately agreed to the process. Only one of the four Gospels gives any detail about the impregnation and the reader gets the impression that she was somehow rendered unconscious during the event.

Since it would be impossible for the cells in Jesus' body to function without a complete set of 46 chromosomes and the egg that Mary provided only contained 23 chromosomes, one must conclude that the beings that carried out the process of impregnating her must have provided the missing 23 chromosomes. From the single reference in the Gospels about the impregnation process, I got the distinct impression that it was carried out via some sort of artificial insemination during which her hymeneal membrane remained undamaged.

If artificial insemination accounts for Jesus' miraculous virgin birth, then there may have been more at stake then producing a human free of the original sin which dooms one to a limited lifespan. I am of the opinion that its real purpose was to create a hybrid human; that is, a person who would appear like an ordinary human, yet who would also have advanced paranormal abilities such as telepathy and a healing touch. As Jesus grew into adolescence, these specially inbred abilities would have started to manifest themselves and he would have realized that he was quite different from the other Jewish boys of his age. Around this time, he would probably have learned of his unusual birth from his mother and have begun to take an interest in religious matters.

Perhaps as a juvenile or early adolescent he wandered about in desolate areas near his hometown of Nazareth when he began to receive telepathic calls to be at certain locations at certain times. There he would encounter marvelous beings who would engage him in telepathic communication. He would certainly have believed them to be angels and representatives of God. From them he would have learned that he was sired by the being that had sent the angels to him. They would have told him that in later life he was to conduct a healing ministry throughout Judea and that he was to preach a message of peace and brotherhood to his fellow Jews.

In his late twenties, Jesus would have had further contact with these beings and the message he was to deliver would be refined. Eventually the message he was to deliver was that those who followed his moral and ethical teachings and believed him to be the son of God would see the eventual arrival *on earth* of a new world wherein death had been eliminated. It would be a world whose living people would have actual contact with the one who had sent the angels to oversee Jesus' ministry. These angelic beings would also have promised to stay in close telepathic contact with Jesus during his life and to intervene and perform occasional miracles for him at a distance from him in order to gather awed followers to his ministry. He need not worry about his assistants being discovered because, aside from having the power of flight, they were also capable of invisibility!

In his late twenties Jesus formally began his ministry when he was baptized by John the Baptist in the river Jordan. A miraculous dove (or something which appeared to be a dove) landed on Jesus and a strong telepathic signal issued from the sky to announce to witnesses that Jesus was a very special individual. The telepathic signal was so strong that each witness's mind "heard" it in his or her own language.

Over the next few years Jesus traveled around the small towns of Judea with his apostles, preaching the message of personal salvation through belief in him, gave moral teachings, and performed many miracles which were mostly healings. At night, he established telepathic communication via prayer with the beings who were monitoring and assisting him with his activities. Mostly, they remained unseen, but told him in what directions he should travel by day. This assured that they would be at those locations in order to assist him with the various miracles in the event that these required additional abilities which exceeded those that Jesus genetically inherited at his conception.

Unfortunately, my research has indicated that the process of telepathic communication is not a perfect one. It is quite possible that during the years of his ministry on Earth, Jesus may have misinterpreted the various telepathic communications that his handlers were transmitting to him at times. Most likely, his misinterpretations only involved certain parts of the message he was directed to deliver to his fellow Jews.

My past study of paranormal phenomena convinced me that telepathy is actually conducted from one mind to another mind through the use of their ocular systems. Thus, a sender's mind causes his mental image associated with a particular word in *his* language to be converted into a complex pattern of tremors in the rectus muscles that control the movements of his eyeballs. These tremors then cause his eyes' rectus muscles to emit a complex pattern of bursts of extreme low frequency or ELF electromagnetic radiation (note that each burst is a just a short train of electromagnetic oscillations with a more or less fixed frequency that issue from a single rectus muscle) which then travel toward a receiver's eyes. The fluids in the eyeballs of the receiver then refract or bend the incoming ELF bursts and thereby focus them onto the receiver's corresponding rectus muscles were they generate nerve impulses that travel into the receiver's visual cortex. Depending on the strength of the incoming telepathic signal, the receiver may "see" an image of the transmitted concept in his "mind's eye" or both see it and subvocalize the word(s) for the concept in his own language and then actually "hear" the word(s) in his "mind's ear" with the "voice" being his own.

While praying, Jesus would have received communication in the form of a powerful telepathic signal from his angelic handlers that, like the incident that occurred during his baptism, originated from the sky, or, more likely, from some sort of airborne craft. The mental images and their associated words that he received would then have been automatically and instantly translated into either Hebrew or Aramaic (which is the language he would have regularly spoken) by his own mind. It would be during this complex telepathic communication process that the possibility for a misinterpretation by Jesus' mind could have occurred.

For example, when Jesus tells his disciples that the kingdom of God is within them, he may have been misinterpreting a telepathic message that was meant to tell humanity that it had the power through its own efforts to achieve a paradise on Earth. When he spoke of a battle of Armageddon that would precede the establishment of God's kingdom on Earth, perhaps the real message was that mankind would have to struggle against and overcome the many problems of the world before its constantly advancing science and technology would allow a paradise to be established on Earth.

When Jesus spoke about his return to Earth before the last apostle died, the actual message may have been referring to future plans by his handlers to return him to the Earth to promote his message in other parts of the world. This message somehow got mixed in with the misinterpreted message he received about a cataclysmic battle of Armageddon that would involve the Roman Empire in the Middle

East. I am of the belief that, after the resurrection of Jesus, he did, in fact, appear centuries later in various other parts of the world including India, North and South America, and even Africa. Many villages in these parts of the world have legends of a light-skinned man with a beard who appeared amongst them one day and provided the sick with miracle cures and people with moral teachings and prophecies.

When Jesus promised his contemporary Jewish followers eternal physical life in an earthly paradise, he may have misinterpreted a telepathic message stating that *eventually* mankind would, through the development of its science and technology, manage to indefinitely extend human life and even "raise" the dead. This, however, would not be happening during the lives of any of his contemporary followers because they did not yet possess the technology necessary to achieve it.

These various misinterpretations, of course, would account for the obvious failure of the central prophecy of Christianity.

Despite these possible misinterpretations of the messages Jesus was supposed to deliver to mankind, there can be little doubt as to the validity and strength of the moral message he delivered which is embodied in his "Sermon on the Mount". Here we see some of the finest teachings on the ethics of human interactions ever delivered. He shows that the quickest path to an earthly paradise wherein poverty, disease, and death are banished requires that humans practice patience, tolerance, forgiveness, and genuine love for each other. Humans are to perceive each other as brothers and sisters in one single global community. Their emphasis should be away from material things and toward helping each other. True happiness and fulfillment will not come through the acquisition of material things, but rather through the possession of nurturing and rewarding relationships with the people around oneself. This simple message is as true today as when Jesus delivered it millennia ago.

It is interesting to note that Jesus lived during a time when there was unprecedented violence and bloodshed in the world. The Roman Empire was in a state of almost continuous warfare with its neighbors and poverty and desperation for most were widespread. The beings who created Jesus and directed his words and actions could hardly have failed to notice these terrible global conditions as they monitored our planet.

I believe that those who directed Jesus were less concerned about the personal salvation and immortality of Jesus' contemporaries and far more interested in the salvation of humanity as a whole on planet

Earth. Jesus and Christianity may only have been one of dozens (or even hundreds!) of religions that these beings sowed on Earth during the long and fragile history of the human race. Perhaps these angelic beings appeared at crucial times in our history whenever the level of violence and murder reached a crisis stage at which point it might actually become possible for all of humanity to destroy itself. In each case they would either select or even artificially create a genetically superior person who would become a telepathically linked conduit through which they would deliver a message of salvation hope to humankind. To change the course of the self-destructive path humanity was on would always require a change of heart that would have to be spread throughout as much of the involved local populations as fast as possible. This change of mentality would require individuals to focus on the important species preserving issues of life and away from such species destroying things as greed, envy, intolerance, and aggression.

Jesus' story ends rather abruptly in the Gospels. He has the one witnessed meeting with two mysterious angelic beings, informed the apostles of his impending execution, and then promised that he would be resurrected as a sign that he truly was in contact with God.

Upon arrival in Jerusalem, he celebrated a final Passover sadir (known in Christianity as "The Last Supper") with the apostles and dispatched Judas Iscariot (the group's treasurer) to lead the Temple guards of the Sanhedrin to Jesus so that they can take him prisoner. After a secretive nighttime trial by the Sanhedrin from which those few high priests sympathetic to Jesus are purposely excluded, Jesus is found guilty of the sins of blasphemy and heresy by them for claiming to be the son of God and then taken first to King Herod (the son of the corrupt king "Herod the Great" who died shortly after Jesus was born) and finally to the Roman Governor of Judea, Pontius Pilate, to be tried and punished if found guilty of the secular charges of sedition and treason. The penalty for being found guilty of either of these Roman crimes was almost always immediate execution!

Jesus offered no defense of himself and Pilate decided to let a mob determine whether Jesus or a condemned murderer, Barabbas (whose first name was also Jesus!), should be set free in observance of a local holiday. The mob, knowing that Jesus was a man of peace and that Barabbas might lead a revolt against their hated Roman occupiers, shouted that Barabbas be freed. This finally doomed Jesus to execution by crucifixion. After a scourging, he was forced to carry the heavy horizontal beam of his cross through the streets of Jerusalem to a hilltop east of the city's gates (the vertical pieces, being too heavy, were kept and then reused at the execution site). He was crucified there along with two thieves that day. After several hours on the cross, he died and, before sunset, was placed into a borrowed tomb.

Three days later on a Sunday morning, a female disciple of Jesus, Mary Magdalene, visited the tomb (some of the Gospels state she was accompanied by another "Mary" who may have been Jesus' mother, while the Gospel of John implies she was alone). The massive stone that sealed its entry had been rolled aside and she encountered a mysterious being, an angel, inside the tomb who told her that Jesus had risen from the dead! Apparently, the Roman guards ordered to guard the tomb were somehow rendered temporarily unconscious during the time (probably early Sunday morning) when Jesus' body was removed. For forty days after his resurrection, Jesus appeared many times to the apostles and, perhaps, to as many as five hundred other people! He finally told the apostles to meet him on the shores of the Sea of Galilee.

At Galilee, the original twelve apostles received their final command from Jesus. They were to take his teachings to all parts of Judea and await his imminent return. He then ascended into a cloud and disappeared from the view of the apostles. The remaining books of the New Testament are somewhat anticlimactic and go on to document the efforts of the early Christians to spread their faith not only throughout Judea, but also the entire Mediterranean area.

The reader should know that the presently accepted books of the New Testament found in the King James version of the Bible are only a fraction of the early Christian writings currently in existence. They were actually *voted* into the New Testament at the Council of Nicea which took place in what is now Turkey in 325 A.D. The hundreds of bishops that attended this large meeting had many biases that influenced their votes on which writings would be considered "canonical" or truly legitimate. One of their goals was to make Jesus appear to be a single, celibate man who was at odds with all things Jewish. However, there are other Gnostic Gospels that were in use in Egypt in the first century AD that suggest that Jesus may have had a romantic relationship with one of his female disciples named Mary Magdalene from whom Jesus earlier exorcised seven demonic spirits. It is even possible that he may have sired children by her! Interestingly enough, Mary Magdalene is the only disciple that is mentioned as being at the foot of the cross during the crucifixion in all four of the synoptic gospels. She is also the first of the disciples to discover that Jesus had risen from the dead.

The reality was that Jesus, although only having one human parent, was raised as a Jew and participated in all of the rituals and customs of Judaism. He says in the gospels, "I come not to change, but to fulfill". Thus, his initial goal in delivering his message to his local Jewish contemporaries was to reform *their* Judaism and *not* to start the new religion of Christianity. Only late in his ministry does this goal seem to change when he charges the apostles to take his teachings to the entire Mediterranean world which is only a very small portion of the Earth's land surfaces. This change, I believe, was necessitated by the undeniable resistance by the vast majority of his contemporary Jewish countrymen to accept his various reforms.

There was an early Christian text that was suppressed during the first century which is known as the "Apochraphon Jacobi". It is very interesting because it gives a slightly different version of the life of Jesus that is portrayed in our present Bible. It contains the usual miraculous healings and moral teachings, most of which appear in our present day New Testament Gospels, but then gives a rather fascinating *different* version of Jesus' ascension.

In the Apochraphon Jacobi, Jesus' final meeting with the apostles takes place on the Mount of Olives which is just outside of and east of Jerusalem. He again charges his followers to love each other and work tirelessly to promote his story and teachings throughout the Mediterranean world. At this point in the narrative, however, a strange dark cloud slowly moves into position just above the Mount of Olives and it begins to thunder and lightning. The apostles are very afraid, but Jesus tells them not to be because it is only God coming to take him away! A flaming chariot (without horses) then descends from the cloud and lands next to Jesus. A door or hatch in the side of the chariot opens and the apostles can make out two figures inside that beckon to Jesus. He enters the chariot, the door closes, and this *conveyance* then begins glowing as it carries Jesus up into the dark cloud. The cloud and Jesus then drift off toward the horizon as the apostles are left behind rejoicing at the miracle they have just beheld. Their belief in the authenticity of Jesus is then given final and overwhelming verification.

From studying both the Old and New Testaments, it became apparent to me that this literature is an abundant source of paranormal phenomena which, as I suggested in my first book dealing with the paranormal, *The Physics of the Paranormal,* can be rationalized in terms of advanced physical principles. This led me to conclude that the events of the Bible are more super technological in nature rather than supernatural. In fact, I would venture to go so far as to suggest that both Judaism and Christianity have one essential detail in common: they are both the result of advanced beings of an *extraterrestrial* origin interacting with ancient peoples. If this is the case, then the teachings of these great religions are certainly compatible with the concepts of the modern New Age movement which places an accent on such topics as spirituality, brotherhood, paranormal phenomena, and, of course, the subjects of ufology and extraterrestrial life.

Many "New Agers" look forward to a day when Earth's people, after developing their own UFO technology, will be welcomed into a local organization of spacefaring extraterrestrial races. This is compatible with the (telepathically misinterpreted) prophecy of Jesus that the "kingdom" of God would come to Earth during the lives of some of his original apostles. Jesus further claimed that after his Second Coming, his disciples who had adhered to his teachings would also be able to perform all of the same miracles that he had done. New Age research into parapsychology and psychic phenomena also promises to one day allow ordinary people to duplicate the various paranormal effects (including healing and even resurrection of the dead) through the use of advanced technological means.

So, in light of the above, I now see Jesus and all the other founders and prophets of other religions as unique people who were either selected after birth or created prior to gestation by extraterrestrial forces for a special purpose. This purpose had two parts. The first was to assure the survival of humanity and the second was to teach it spiritual values that would eventually help prepare it to take its rightful place in the community of sentient lifeforms within our galaxy. This process has, most likely, been ongoing for tens of thousands of years and the religion of Christianity is actually only one of the more recent consequences of it. Jesus was a sincere man who did his best to perform as required by his extraterrestrial sponsors even though it caused him to suffer an excruciating form of execution.

As to the present status of Jesus, I can only offer the most general of speculations. I believe that he did come back from the dead and then went on to continue spreading his message in other parts of our planet at other times. As I have showed in some of my other writings, the technological ability to resurrect dead organisms also implies the ability to bestow immortality upon them. Therefore, it is entirely theoretically possible that the person the western world knows as Jesus could, at this very moment, still be alive somewhere out there in our galaxy. Perhaps someday soon, after we of Earth make "official" contact with extraterrestrial beings, we will see Jesus again. At that time we can look forward to a complete explanation of what occurred during his lifetime in Judea and exactly what the motivations and intentions of our cosmic neighbors toward humanity were at the time he was born.

To conclude this first section of this chapter, I need to briefly mention another topic.

Assuming that one can accept my hypothetical version of the Gospels given above, one might wonder why extraterrestrials would promote the existence of a supreme God in their interactions with mankind. Would it not have been much easier to just promote brotherhood among humans *without* introducing the concept of a deity? For example, the "angels" who contacted various ancients could have just stated that they were flesh and blood creatures like humans that depended upon technology to achieve their various "miracles". Surely, this would so have impressed ancient people that they would all have eagerly embraced subjects like mathematics, philosophy, etc. and began to rapidly develop Earth's science and technology. Why the need for angels, deities, miracles, and prophecies?

Upon giving the matter some thought, I realized that there are several reasons why this honest approach was not used. Primarily, ancient people had little or no concept of technology. They basically lived in a "magical" world that they imagined was controlled by spirit beings. To get their attention, an extraterrestrial being or "angel" would have to claim that he represented a supreme spirit who was more powerful than all of the rest. To prevent general panic among ancient humans that would only have made them scatter in all directions, the angelic beings would always limit their contact to only a carefully selected individual and his immediate associates. Since primitive people usually only believe what they can directly sense, it is important that immediate prophecies be fulfilled more or less as stated. It is also important that, after a core of followers has been secured, the person selected for contact by these extraterrestrial beings is removed from the scene. This lends an air of mystery to the leader's life and prompts his followers to spend their future lives disseminating his message (which is really the message of our local spacefaring extraterrestrials) as far as possible so as to grow the total number of followers.

As for promoting belief in a supreme being or God, there is also a simple explanation for this. It may express a fundamental belief of all extraterrestrial races that there is, in fact, a supreme self aware intelligence to our cosmos. I, of course, do not believe that this being takes the form of a Zeus-like character found in most ancient religions which imagine God as a vengeful, bearded giant who dwells in a cloud enshrouded heaven above waiting to hurl lightening bolts down at evil people. Rather, the "God" of the extraterrestrials is actually a fundamental intelligence that exists throughout the infinite cosmos at a *subatomic* level. It is this intelligence which maintains the stability of physical reality and is responsible for the emergence of all lifeforms wherever the conditions for their existence are suitable.

This universal cosmic mind is simultaneously aware of all creation and always functions to bring forth balance, order, and harmony. While a humanoid brain must depend upon neurons and chemical reactions to be aware of what is happening in the body that carries it, the cosmic mind depends upon various faster than light exchanges of information taking place between all of its infinitely dispersed subatomic particles. Because the complexity of this infinite number of interactions infinitely exceeds what occurs within a single living creature's organic brain, we can expect the cosmic mind's level of awareness and control to far surpass what any single brain is capable of ever achieving.

Quite possibly, this supreme cosmic intelligence responds to the well intentioned needs of its "children" (that is, all sentient lifeforms throughout the cosmos) by subtly changing their realities to meet those needs. It must do this in order to maintain the balance of the cosmos. Occasionally, it even allows for "miracles" of a highly improbable, but not quite physically impossible, nature to take place that the secular world tends automatically dismiss as nothing more than random chance. But, this supreme intelligence is omnipresent and accessible at any time of the day or night to those who believe in it. You can always reach it with your prayers.

The above concludes the "pro-Christian belief" first section of this chapter and what follows is a far more negative "anti-Christian belief" second section whose conclusions I currently accept as being closer to the truth of the matter when it comes to Christianity. Again, these are only my personal beliefs and the reader should draw his or her own conclusions based upon consideration of the material of this chapter as well as that available from other sources including their own personal experiences.

Sometime around Easter of 2007 a very interesting Canadian documentary film was shown on one of my cable television networks. It was about a film crew that had gone to Israel to investigate an ancient Jewish burial tomb that had been found a few miles south of "The Old City" in East Jerusalem in an area called "Talpiot". During some construction work being done there in March of the year 1980, the entrance to the underground tomb had accidentally been exposed. Once the discovery was made, all construction ceased for a few days and the Israeli Department of Antiquities, which is in charge of preserving archaeological sites, was notified. They entered the tomb and found it to contain 10 limestone boxes or "ossuaries" that contained the skeletal remains of a group of deceased Jews from the first century AD and some of the boxes may have contained more than the remains of a single body. Several of these boxes also had "epigraphs" or names and phrases scratched into their soft exterior surfaces that were in various languages used in the first century AD in Judea. However, after various scientists examined the skeletal remains for a few days, a group of rabbis were then allowed to take them away for burial as is the current religious custom in that country.

In the first century AD, the Jews adhered to burial ritual that involved anointing the newly deceased body with various oils and spices and carefully wrapping it in a shroud that was then tied to the body at various points. The enshrouded body was finally placed on a table in the tomb and allowed to decay over the course of a year. After that time had passed, there would be virtually no flesh left on the deceased's bones and the tomb would be reopened, entered, and the bones then placed into a small limestone ossuary box with a lid. Often, the deceased person's name, familial relationship, and title if any would be scratched onto one of the larger side surfaces of the soft limestone box and it would then be placed into a niche or shelf carved into one of the walls of the tombs. An average sized tomb could easily hold the remains of a dozen or more bodies.

At first glance this tomb found at Talpiot seemed not that different from about 1000 other similar tombs that had previously been found in the "Holy Land".

Apparently, the documentary film makers had found out about this tomb because earlier in 2002 one of the ossuaries rumored to be from his particular tomb had suddenly made a public appearance. It had been privately purchased from an anonymous seller and began being put on display in both Europe and the Canada. That limestone box had Aramaic writing on it that read "James, brother of Jesus"! That was enough to provoke the interest of the filmmakers and they arrived at the site only to find that the entrance to the tomb had been covered over with a thick slab of concrete which had been installed because local children had been found playing inside the tomb after its ossuaries had been removed.

As the filmmakers' hired workers tried to lift the security slab, the local authorities were summoned and they were prevented from continuing their efforts so that the tomb could not be physically entered. Even if they had been able to enter it, however, there would have been little to see. All of the original ossuaries had been removed years earlier and were in the possession of the Israeli Museum and had already undergone extensive testing to assure that they were, indeed, from the first century AD. Also, all of the epigraphs on the exteriors of the limestone boxes had been translated. The entire matter had then been kept quiet and basically forgotten about for almost 27 years until the "James Ossuary" suddenly surfaced in 2002.

What really made this particular tomb interesting was that the inscriptions on its ossuaries indicated that it contained practically the entire Christian holy family!

Most notably, one of the ossuaries was inscribed with the Aramaic letters that translated as "Jesus, the son of Joseph". Another was marked, in Hebrew, with the letters that spell "Mary" who had been his mother. Two of brothers that were mentioned in the Gospels were there too.

Their ossuaries read, in Aramaic, "James, the brother of Jesus" and "Jose" which was the nickname used for Jesus' younger brother. Another box was marked, in Aramaic, with the name "Matthew" who had probably been an associate of the holy family.

Then there was a box marked, in Greek, "Mary, the teacher" which was an honorary title given to Mary Magdalene because of her efforts to promote Jesus' teachings. And, finally, a box marked in Aramaic that read "Judah, son of Jesus"!

The box containing Mary Magdalene's bones had been place in the same niche with the one marked with Jesus' name and had been placed on top of his box. This was a placement that was customarily used to indicate that the two people had been married while alive and indicates that Mary Magdalene had been married to Jesus and that Judah had been her son!

Because the bones had been removed from these ossuaries and buried, it was not possible to extract the nuclear DNA from any remaining and intact marrow cells they might contain and then perform an extensive genetic analysis based on that genetic material. However, despite this the inside surfaces of these boxes did have a thin film on them referred to as "patina" which is composed of a unique blend of materials contained in the air of a particular tomb and any mitochondrial DNA that might have been absorbed onto the surface of the limestone from any decaying cells still adhering to the bones when they were originally placed into the boxes. The filmmakers were able to obtain two samples of these residues and an analysis of the genetic material they still contained showed that the residues from the Jesus and the Mary Magdalene ossuaries were *not* biologically related which would be expected if they were a married couple.

Needless to say, when I finished watching this documentary I was stunned by the implications of it and it was obvious why this information had been suppressed for a quarter of a century. If this was, indeed, the tomb of the holy family, then it meant that Jesus was not a celibate bachelor as he has traditionally been made out to be. He had married and had a son. Also, the fact that his bones were found in a tomb indicates that there was no physical resurrection or ascension and, consequently, there will be no future return of Jesus to Earth. Once these Bible "truths" are dismissed, it then becomes much easier to dismiss the rest of the Jesus story.

There have been several other burial sites in the past that made claims to being the tomb of Jesus, but there was something about this one that made it stand out in my mind. Needless to say, its discovery has provoked much controversy in the religious world with various archaeologists and theologians attempting to explain it away or minimize its relevance. One concern is that the tomb contained more than the remains of ten bodies and that some of the ossuaries, therefore, contained more than a single body. Another concern is the use of different languages on the limestone boxes. If they were all members of the same family, then one wonders why all of the boxes were not marked with epigraphs in the same language which, most likely, should have been either Aramaic or Hebrew.

I don't find these questions troubling, however.

Regardless of how many bones may have been placed in any of the ossuaries, the mitochondrial DNA testing of the patinas from the two boxes that were marked as *originally* containing the remains of Jesus and Mary Magdalene showed *no* genetic relationship between *any* of the different mitochondrial genetic materials contained in these two patina samples. Since those two patinas, even if they each also contained several other individuals' mitochondrial DNA, would still have had to have contained the mitochondrial genetic materials from both Jesus and Mary Magdalene and that fact automatically means there was no genetic relationship between the body cells of Jesus and Mary Magdalene.

Some have suggested that the use of different languages in the epigraphs scratched on the limestone boxes indicates that they could not contain the holy family members mentioned above. Again, this proves nothing. We must remember that these ossuaries contained the skeletal remains of people who died at different times and, perhaps, in different locations. The language used for the names on the boxes could have just been the result of the person having died in an area where that language was prevalent or, perhaps, was the one preferred by whoever arranged for the burial.

What finally convinced me that the Talpiot tomb was genuine was a mathematical analysis of the find made by a Canadian professor of mathematics and statistics. He used the names on the boxes along with their frequency of usage in first century AD Judea and determined that the chance of all of those names occurring in that one tomb *by random chance* had a probability of between 1/600 and 1/1,000,000 depending upon how one selected certain variables that were used in the analysis. If we use the most probable chance of the occurrence being due to random chance of 1/600, then that means the probability of their particular names *not* being found together due to random chance was 599/600 or 99.83+ %. That is high enough for me to consider this tomb as being "the" one containing the remains of Jesus of Nazareth along with his mother, brothers, wife, and son.

As I considered this information, I began to research more about various events that took place in the first century AD following the execution of Jesus. From this I was eventually able to piece together what I believe most likely happened back then. Of course, I can not prove this scenario as being 100% certain and I only offer it here for the reader's consideration. As I stated earlier in this chapter, each of us must reach our own conclusions regarding *any* religion based upon our particular experiences and what we believe is reasonable. It is not my purpose here to "convert" anybody to my current beliefs, but, rather, to just document them as they now stand. Also, that does not mean that my opinions may not change in time even though I now feel that is unlikely.

I found it interesting that the virgin birth of Jesus is only mentioned in one of the four Gospels, yet this is supposed to be proof positive that he was the son of God and not Joseph who only served as a foster father. One would think something as important as that would, like the crucifixion and resurrection, have been mentioned in all of the Gospels. Then there is the matter of all of the "brothers and sisters" that Jesus had which are mentioned in one of the Gospels.

The Roman Catholic Church suggests that they were merely children Joseph had earlier with some earlier wife and, thus, Mary could have remained a virgin for the rest of her life after Jesus' birth. That earlier assumed wife, however, is never mentioned in the Gospels. If one accepts that interpretation and Jesus had at least one brother who was younger than he and who was the "Jose" whose name was inscribed on the Talpiot ossuary mentioned above, then there is only one way that Mary could have remained a virgin after Jesus' birth. Either Jose's birth was also the product of immaculate conception that, once again, did not involve Joseph or the latter took another wife after the birth of

Jesus which would only make sense if he divorced Mary or if he was practicing polygamy.

From these details, I eventually came to believe that Jesus was nothing other than 100% human. He was the first and eldest of about seven children. As the first born, he would have had considerable pressure on him to follow in his father's profession which, based on the latest translations, was that of a skilled construction laborer and would have included carpentry. Jesus, when he became a teenager, might have grown weary of the grueling labor involved in such a profession and began to enter into more and more conflict with his father who had decided that he should be happy with this occupation since he expected to take it over so that his aging father could retire some day.

Jesus, however, had different plans. He lived in a time when the Jews of Judea were enduring the harsh occupation by the Romans. The Jews wondered why, if God was with them, that God had allowed the Romans to reduce their country to a vassal state of the Roman Empire and install "Herod the Great", an Arab and not a Jew, as their king. The ultimate insult to Judaism was to have a statue of the Roman god Jupiter (equivalent to the Greek god Zeus) placed right in their Temple at Jerusalem which was the center of religious life in Judea.

This was to the Jews an extreme form of blasphemy, yet their god Jehovah did nothing about it or their Roman occupiers. In response, all that the Jews could do was hope for the arrival of a long prophesized messiah who, like their former kings David and Solomon, would quickly expel the foreign pagan invaders from their country.

As a teenager, Jesus would have been exposed to these historical events and the implications they had for the religion of Judaism. Perhaps it was then that he conceived the idea that the problem was not really the Roman, but the Jews themselves! Maybe, he reasoned, Jehovah was not aiding the Jews because they had somehow failed in their religious duty to him. Eventually, he would have reasoned that the problem was not the average Jew, but, rather, their religious leaders who he came to see as hypocrites more interested in their social status and comfort than in practicing Judaism the way God really wanted them to. To again gain the favor of Jehovah and rid themselves of their Roman conquerors, Judaism would have to be reformed and practiced the "right way" by the Jews of Judea.

When he was not busy helping his father with construction jobs, Jesus would have started to espouse his beliefs to others and found that many of those agreed with him. But, who would lead such an effort to reform Judaism they wondered?

Sometime during his early twenties, Jesus must have had some experience that convinced him that he was the promised messiah of the Jewish people. Perhaps it was an unusually vivid dream, or he saw a "wonder" in the sky such as a meteor or even a UFO, or he developed some sort of obsession with the idea for some psychological reason. We will never know for sure, but the result was that he became firmly convinced that, in reality, he was not the biological son of Joseph and that his "mission" on Earth was, in fact, to reform Judaism. This reform, aside from removing the Romans from their country, would also allow those that practiced it to eventually achieve physical immortality right here on Earth! Most likely, this metaphysical salvation scheme Jesus proposed had been borrowed from the Persian religion of Zoroastrianism with which Jesus would have been familiar. It preached that there was a continuous struggle between good and evil in the world and that there would be a coming "great day of judgment" during which all of the evil people would perish and only those that had been steadfastly good throughout their lives would be saved from annihilation. Those good individuals would then be granted immortality in some sort of afterlife as a reward either on Earth or in some dreamy spiritual realm.

Eventually, the conflict between Jesus and his father Joseph would have reached the point where they would be unable to work together and his next younger brother would have taken his place. At that time Jesus would have left home and began wondering about the countryside with a small group of loyal followers, the apostles who had likewise abandoned their families in order to follow Jesus. They wandered about preaching his message in various synagogues and to large crowds that would gather outside of small towns to hear him preach to them. They also collected small contributions from any who would contribute and that helped pay for their food and shelter.

As he traveled about preaching like many other itinerant rabbi of that time, desperate people, thinking he had some magical healing powers, would occasionally approach him and beg for a healing. In most cases nothing would happen, but, on occasion, someone with a serious health problem would, a few days later, suddenly become well again. There is nothing miraculous about such healings. They would just have been the result of the so-called "placebo effect" in which a person gets better because he believes he will get better. The news of these "healings" would then be spread about as rumors and slowly distorted and amplified over time as they were passed from one person to another by word of mouth. Before long Jesus would be credited with curing the blind and even raising the dead even though these things did not actually happen.

Slowly, over the course of years, Jesus would have developed quite a following including several female admirers. No doubt Mary Magdalene was one of these and it was not long before she became part of Jesus' closest followers. Eventually, they probably married (all of the original apostles were married men) and she bore him a son named Judah whose remains were contained in one of the ten ossuaries at Talpiot.

Sometime around the age of thirty, Jesus concluded that he indeed was the son of God and destined to reform the Judaism of that day and return the favor of Jehovah back to the Jewish people. The size of his following and his constantly speaking out against the high priests who operated the Temple in Jerusalem would immediately have brought him to their attention. Most of the Jews of his day only considered him to be yet another failed messiah who was unable to expel the Roman invaders and the Roman officials in Jerusalem just considered him to be a political nuisance who might incite an uprising against them if he managed to gather enough followers. There had been previous uprising whose suppression had resulted in much bloodshed and even further ill feelings toward the Roman occupiers. However, the high priests themselves saw an even greater threat in the continuing activities of Jesus and his followers. He was actually saying that they were responsible for the plight of the Jewish people and for God apparently turning his back on them! Needless to say, they saw a quick and easy solution to this problem: eliminate Jesus as quickly as possible.

Jesus, after he had entered the Temple in Jerusalem during the preparation for the coming Passover holiday and overturned the money changers tables, knew that he was being sought by the Sanhedrin priests for the purpose of execution. At this point, he is really unconcerned about being captured, tried, and then being found guilty of various charges and executed. The reason was that by this time Jesus actually believed that his real father, God in heaven, would either save him during the execution process or, if he actually died, miraculously bring him back to life again as a final sign that he was the son of God and the new spiritual leader of the Jewish people.

Delusions can be powerful things and can make a person needlessly injure himself or even lose his life if he acts on them. They are particularly dangerous when one is surrounded by a band of loyal followers who reinforce the delusions because they too have come to accept them as reality. Because of this, the apostles would not have tried to talk Jesus out of surrendering himself to the Temple guards who were actively searching for him. Like him, they believed that Jesus would rise from the dead and eventually his teachings would be accepted worldwide (which for them spanned from the Mediterranean area eastward to India) just before his triumphant bodily return or "Second Coming" to temporarily rule over God's kingdom of heaven on Earth for a thousand years until God could personally come down to take over the job.

We in the western world are, of course, well familiar with the events that quickly followed because of their depiction in many religious movies made during the last century.

Jesus celebrates a final Passover sadir and then dispatches Judas Iscariot, the group's treasurer, to lead the Temple guards to the Mount of Olives outside of Jerusalem where Jesus would be waiting for them. Judas then kisses Jesus to identify him so the guards can arrest him.

Jesus is taken for a quick nighttime trial at the Sanhedrin at which he is found guilty of blasphemy and heresy. He is then taken to King Herod for whom he fails to produce any miracles and who then decides to send him to Pontius Pilate, the military governor of Judea at the time. Before Pilate, Jesus offers no defense and Pilate decides he is guilty of sedition and treason both being crimes which carry a death penalty by crucifixion.

However in observance of a local holiday, Pilate decides to let a mob outside of his fortress overlooking the Temple decide whether Jesus bar Joseph (that is, Jesus the son of Joseph) or Jesus Barabbas, a rebel leader convicted of murdering a Roman soldier, should be spared. The crowd mostly shouts for Barabbas to be freed and so he is while Jesus is prepared for crucifixion. Apparently, the crowd wanted Barabbas because he had previously led a riot against the Roman occupiers in Jerusalem and they hoped he would continue to attack the Romans.

After a brutal scourging and mockery by his guards, Jesus is made to drag the horizontal beam of his cross through the streets of Jerusalem to an execution site outside of the walled city. It is a long and arduous trip and Jesus is then, after his Roman soldier executioners assemble his cross, crucified and his cross' upright post is set into a hole in the ground between those of two other convicts that were being crucified that same day for thievery.

As he hangs slowly dying on the cross from respiratory distress and loss of blood, Jesus says several things. One quote found in the Gospels is that he says, in Aramaic, "Father, why have you abandoned me?" He was apparently surprised that there had been no intervention by God before he reached the cross and at this time he would have realized that he was actually going to die in only a matter of a few hours of time.

In all four Gospels Mary Magdalene is mentioned as being at the foot of Jesus' cross as he is dying. Again, this tends to reinforce the idea that he and she had a special relationship as would have existed between a husband and a wife. Another Jesus quote has him saying "Son behold your mother." It is generally believed that Jesus was addressing the apostle John and telling him that he should treat Jesus' mother Mary as he would his own mother now that Jesus was about to die and would no longer be around to protect and provide for her in her old age. There is an oral tradition that states that John, the only apostle that did not to die a martyr's death, did take care of Jesus' mother after he was finally released from the isle of Patmos in the Mediterranean Sea to which the Romans had exiled him. Other oral traditions have John care for her in the Greek city of Ephesus where she was eventually buried. The Roman Catholic Church maintains, however, that Mary never actually died, but, rather, was "assumed" into heaven by Jesus which, obviously, would be a false assumption if her remains were contained in one of the Talpiot ossuaries as, apparently, they were.

Considering the discovery at Talpiot and other reliable sources, I now believe that Jesus was not speaking to the apostle John from the cross, but, rather, to his *son* Judah and the Mary that he, Judah, was suppose to behold as his mother was none other than Mary Magdalene! In other words, the correct interpretation of this Gospel quote is that Jesus was telling his biological son that, now that his father, Jesus, would soon be dead, it was up to Judah to care for his aging mother Mary Magdalene. If this was the case, then one wonders why Jesus would be concerned about his wife's future welfare if he would be resurrected from the dead by God. Jesus had already wondered why God had not intervened on his way to the execution site. Maybe Jesus, with this remark to his son, was also beginning to have doubts about having a future resurrection as well?

There is a historical source that says that in the year 42 AD a small boat without sails arrived on the southern Mediterranean coast of France (then called Gaul). There were several people in this boat that included two women named Mary and a 12 year old girl named Sarah. I believe that those Mary's were Jesus' wife, Mary Magdalene, and his mother, Mary. The young girl was probably the daughter of Jesus by Mary Magdalene. Sarah in Hebrew means "princess" and that would have been a fitting name for the daughter of the "Prince of Peace" or Jesus. This daughter and her mother Mary Magdalene stayed in southern France for an extended time. Eventually, the two Mary's returned to the Middle East while the grown up daughter stayed in France and married into a noble family there which explains why her ossuary was not found at the Talpiot tomb. Interestingly enough, there was a basilica eventually built near where their boat landed that was dedicated to Mary Magdalene who is considered the patron saint of prostitutes. Many Christians believe that she was the prostitute that Jesus saved from being stoned to death by a mob ("Let he who is without sin cast the first stone."), but there is no evidence in the Gospels that Mary Magdalene was that woman who Jesus saved. Supposedly, Jesus' first encounter with Mary Magdalene was when he exorcised seven demons from her, but, again, she could have just had some emotional problems that were corrected by the placebo effect of believing that the messiah had healed her.

With this additional information, it became obvious to me from the age of Sarah when she arrived in France that Mary Magdalene must have been pregnant with her on the day of Jesus' crucifixion! But, why, I wondered, would Sarah and her mother suddenly appear in France 12 years after the execution of Jesus and where was his son Judah?

Actually, there is a simple scenario that explains all of these details and further reinforces the validity of the Talpiot tomb.

After Jesus' execution, his body was wrapped up and placed into the borrowed tomb as described in the Gospels. Pilate did post guards at the tomb to make sure no attempts were made to steal the body before *parts* of at least three days had passed. During this time, the apostles and Jesus' pregnant wife Mary Magdalene and her teenage son waited impatiently for the time to pass so that they could again be with their leader who they fully expected to rise from the dead.

The ancient Jewish day actually began at sundown and lasted until the next sundown. So, if Jesus died at about 3 pm on a Friday afternoon and was placed into the borrowed tomb *before* sundown of that day, then he would have been dead for only a small *part* of a Jewish day. From sundown Friday until sundown Saturday he would have spent an entire 24 hour Jewish day in the tomb. And from sundown Saturday until sunrise Sunday he would have spent about half of a Jewish a day in the tomb. Although that only adds up to a few hours more than 36 hours, technically, it does at least include some of each of the three days that he prophesied that he would be in a tomb before he was miraculously raised from the dead.

At sunrise the following Sunday after Jesus' execution, the guards would have gotten tired of guarding the tomb of a dead man and gotten permission to leave their post there. At that time Mary Magdalene, unable to wait any longer, would have arrived and found the guards gone and the tomb still sealed. Sharing Jesus' delusion about being resurrected she would have been desperate to have the huge disc-like stone rolled away from the tomb's entrance so that her husband could emerge and, once again, be with her. Perhaps she found some passing workmen and offered to pay them well for the few minutes of effort that would be required to unseal the tomb's entrance.

As she entered his opened tomb, she would have quickly realized that he was *still* dead. That meant that he was not who he claimed he had been: a messiah, directly descended from the Jewish Jehovah, who would reform Judaism and eventually free the Jews of their Roman invaders. The shock of having all of her delusions suddenly evaporate after having reinforced them for years must have been enormous for her.

She also had an additional problem. Without Jesus and his alleged miracles, their small band of followers had nothing to attract the crowds eager for miracle cures and their financial contributions that helped pay for the food and other expenses of the group. Mary Magdalene was now a widow with a teenage son and child on the way to support. Life for widows in the ancient world was harsh. As today, some would be forced into prostitution in order to afford food and shelter if they did not have a family to take them in.

It was at this point, I believe, that she came up with a brilliant solution to the reality that her husband Jesus was now dead and most likely to remain that way. Perhaps she had a friend in the Talpiot area south of Jerusalem that had a large family tomb that was not yet full. Perhaps this friend owed her a favor for some reason. She then unwrapped Jesus' corpse and left the single sheet that had enshrouded his body neatly folded up inside of his borrowed tomb. That would become Shroud of Turin which is now in possession of the Roman Catholic Church. Next, she managed to get one of the passing workmen to remove the body and take it by animal drawn cart down to the second tomb at Talpiot where it was wrapped in a fresh shroud and laid out for a year as was the Jewish custom before the remaining bones were placed into a limestone ossuary that was marked with the decease's name and familial relationship.

Since none of the apostles had seen any of this happen and were still expecting a miraculous resurrection, Mary Magdalene decided to give them one. After Jesus' body was safely out of sight, she would have returned to where the group was hiding out after the crucifixion and breathlessly announced that she had gone to the borrowed tomb and that it was empty! Then she would have said that Jesus had suddenly appeared, approached her, and announced that he would be visiting the apostles soon. Needless to say, the apostles would have run out to the borrowed tomb immediately to confirm her story and, seeing the massive stone cover rolled aside and the tomb empty, would have believed that their master had, in fact, risen from the dead. They would have been ecstatic and could then use this final miracle to promote belief in Jesus' authenticity to others. As long as Mary Magdalene did not reveal the hoax she had perpetrated, all would be well and she need not fear for her and her children's financial future.

If Jesus' son Judah had been a teenager when his father was crucified in 30 or 33 AD, then Judah would have been in his mid to late 20's in the year 42 AD. As the son of Jesus, Judah would have been his heir and the logical choice to lead the early Christians. However, I believe that somehow Judah was murdered to remove him from the scene. Either certain members of the Sanhedrin, still feeling threatened by the growing Christian cult decided to do the deed or, perhaps, someone inside the cult itself decided that one messiah was enough and arranged for Judah's death. We'll never know for sure unless new archaeological evidence surfaces or the bones of Judah can be located and examined to see what kind of death he died and his approximate age at the time of death. For example, if his head had been traumatically severed from his body, then, most likely, he was the victim of some sort of official execution. Severe damage to the ribs or spinal column might indicate that he had been run through with a spear or sword and would point toward an assassination by those who knew his movements.

After Judah's death, his mother Mary Magdalene would have been desperate to prevent the same thing from happening to her young daughter Sarah. After consulting with Jesus' still living mother Mary, Mary Magdalene would have been given some sage advice. Jesus' mother would have told her to get out of Jerusalem and even Judea itself. I think the two older women would have decided to flee over to Egypt and would have traveled by camel along the southern shores of the Mediterranean Sea. However, being pursued by assassins paid to kill them, the women and Sarah kept moving until they reached present day Morocco. There a decision was made to take a boat up along the eastern coast of Spain and, finally, they and several other passengers were put ashore in southern France where they finally found refuge.

With the two Mary's out of the way and no additional heirs around, the apostles were free to promote Jesus' teaching or, at least, their versions of it to the Mediterranean world. They, of course, endured much persecution in the process. But, as the cult spread slowly, the money being taken in would have also begun to grow. Lower level members of "true believers" would have been living mostly on handouts from sympathetic listeners that they were trying to "save". The upper level members, however, would have been enjoying a far more comfortable existence and would be determined to maintain that status quo. Anyone who challenged their authority in deciding what was "holy" and what was not would be branded a heretic and expelled or worse.

Slowly, over the course of several centuries, Christianity, like all man made religions, became a business. The product it sells for member's money or assets is actually an invisible one and really a state of mind. Those that follow a particular denomination's rules (there are currently thousands of Christian denominations each of which believes that it is the one practicing Christianity the "right way") can consider themselves forgiven for all of their various behaviors and thoughts that are offensive to God and would surely bring them punishment in this life or some imagined afterlife. They are further assured that, while their sins are real, death is not real and they can surely escape the obvious finality of it by simply practicing the various beliefs and rituals of the particular denomination. Religions, in general, do not like members engaging in critical thinking. A religion's leaders have already done that so there is really no need for the lower level members to do so.

This, basically, is my current belief about the Jesus story. I believe that he was an actual historical person who, becoming obsessed with the religion and politics of his country, came to believe that he had been chosen by Jehovah to set things right by reforming Judaism which would then reform his fellow Jews. He attracted a small group of followers and, together, they began to reinforce each other's delusional beliefs. As he began to act on those beliefs, he inevitably came into conflict with the powerful religious authorities of his country and that resulted in him being executed. Once dead, he stayed dead and his body was quietly moved to another location so that his followers could be presented with an apparent resurrection. When he did not appear to them later, some reason was made up to rationalize this. Perhaps Mary Magdalene claimed she was having further meetings with her husband Jesus and was delivering messages to the rest of the apostles to direct their efforts. In time she would have come into conflict with the apostle Peter who could claim that Jesus himself had appointed him to lead the early church and not Mary Magdalene when the messiah said to him "Thou art Peter and upon this rock I shall build my church." Maybe Peter was even somehow responsible for Mary Magdalene having to flee from Jerusalem in 42 AD.

Only a few years ago, I learned another surprising fact that made me doubt the veracity of most of the Gospel accounts of Jesus.

There was an interesting anti-religious documentary titled "Religilous" that was hosted by a television comedian and was available for free viewing for a while from my cable television provider. In this documentary, they brought up a very interesting and little known fact which was that the Jesus story was *not* new in the first century AD! There were various other *earlier* versions of it involving others from the region around the Middle East.

For example, the documentary mentioned how in the year 1280 BC a papyrus scroll was published in ancient Egypt. This scroll told about the life of their god Horus.

According to the scroll, Horus was the son of two other Egyptian gods, his mother Isis and his father Osiris. Horus, like Jesus, was born to a virgin mother. As Horus grew up, he began to perform various miracles. He healed the sick, cast out demons, raised the dead, and even walked on water! He eventually attracted 12 followers that became his disciples and proclaimed that he was truly a god throughout Egypt. Horus had himself baptized in a river by a man named Anup the Baptizer. Later, Anup was captured and then killed by beheading.

After a while Horus and his 12 disciples attracted the attention of various priests who decided they represented a threat to the established Egyptian religion. Horus was the first of the 13 killed and his body was placed into a cave. Three days later two women arrived at the cave and it was empty. They then proclaimed that Horus had risen from the dead and was the light of the world and the hope of mankind.

Needless to say, after learning of this earlier Jesus type story, I was struck by how very similar it was to the version involving Jesus of Nazareth that is found in the Gospels of the New Testament. I realized that the authors of the New Testament versions must have had some accurate details about a *few* incidents in the life of the real Jesus and then simply combined them with those found in the scroll of Horus to create what appears to be an accurate historical account of Jesus. Because of this I think it is only safe to assume that the description of the arrest and execution of Jesus is accurate, while his alleged miracles, resurrection, and final ascension are purely fictional. Other incidents in his life which are described are probably a mixture of fiction and truth, but it will never be possible to know what the percentages of each are in any particular incident.

So, the preceding now concludes the "anti-Christianity" section of this chapter, but I want to finally conclude the chapter with a few more comments on religions *in general.*

I think the major problem I have with all religions is that they require their followers to accept a somewhat fixed and simplistic view of reality. The leaders in a religion decide what is real and not real and what is good and evil based upon their interpretations of ancient and mostly fictitious "holy" books combined with, perhaps, their special "revelations" from God himself. Then, in order to be a member in good standing, aside from making regular financial contributions, one must mindlessly accept that religion's particular view of reality. Often, especially now in the technologically advanced 21st century, this "reality" can almost immediately be seen by any sensible person to be in conflict with established scientific facts and to needlessly fill one with shame for having completely *normal* human needs and weaknesses. Usually members are required to suspend their critical thinking faculties and just embrace whatever collection of delusions a particular religion "sells" to them.

Most religions, convinced that only they have the truth, tend to make their members suspicious of "outsiders" (also known as pagans, heretics, infidels, etc.) who, because of an accident of the location of their birth, happen to be practicing a different religion. This inherent paranoia occasionally escalates to the point of violence and even widespread murder. Someone once noted that most of the wars in history have been fueled by religious differences as each side tried to wipe out the other in order to "prove" that God was really on *their* side and favored their victory. Even though many of the world's major religions got started thousands of years ago, it's incredible to see that even today various religious differences around our war torn globe are still causing people to slaughter each other in the name of the *same* God!

Why, in light of all of these deaths, are religions still tolerated in the modern world? The answer to this question is not easy to provide because it involves many issues, but I will try to give some opinions on the matter.

First and foremost, people are scared of physical death. They are not particularly afraid of the *process* of dying because about 75% of all people who die will spend the last three days of their lives in a coma and will not even be aware that they are dying or have died. What really frightens them is the finality of death; that is, *staying* dead once they die. Once dead, they will never again see the sun rise, get together with family and friends for a party, enjoy fine food and drink, or experience sexual pleasure. And, without a functioning brain, they will not even know that they are dead which is probably a good thing considering what can happen to an unembalmed body that is placed into the ground for burial.

Of course, all of this only happens *if* there is *no* afterlife or some vague spiritual realm where a person's soul or invisible animating spirit can live on and still have his consciousness, memories, and, perhaps, the hope of one day inhabiting another living body for another physical life on Earth or some other Earth-like planet. Religions, having nothing tangible to sell, all immediately step in and claim that they can help their members in good standing obtain such an afterlife. Once that is secured, the person need not worry again about what will happen to him after death. This at first might seem like a great relief, but, as one grows older and approaches his own actual death, he may find himself starting to have some doubts about whatever metaphysical salvation scheme he has bought into.

These doubts are the result of one never, during his lifetime, having seen any deceased person experience a resurrection, reincarnation, or any other sort of return from a supposed spiritual realm. One might well wonder why, if the promised afterlives religions sell are in fact real, one's previously deceased loved ones do not occasionally pay a visit from their afterlife to let him know how they are doing "up there" or even "down there"? With the absence of such non hallucinatory visitations, the doubts can linger and grow and that probably accounts for the growing numbers of agnostics and atheists around the world today.

To be convinced in this cynical age, people need to have personal experiences that convince them something is physically real. Far more people have seen UFOs than angels or the spirits of their departed loved ones, so that probably accounts for why about 75% of the western world's populations currently believe in the existence of UFOs and the extraterrestrial humanoid life that will have constructed them and which are operating them in our planet's skies.

Next, religions tend to try to make their members believe that they teach moral values that help humanity live in peace and harmony with each other. Unfortunately, most religions do not bear too close a scrutiny when it comes to moral issues. Their histories are filled with corruption, perversion, rivalry, and violence. As mentioned above, most of the wars fought in history have their roots in religious differences that were the result of religious "moral" teachers preaching intolerance and even hatred for "outsiders" who refused to convert to their particular view of reality. The divisiveness this leads to can only lead to a less peaceful world filled with strife and unnecessary deaths.

Finally, when it comes to answering the question of why religions are still tolerated, one needs to realize that the western world's political systems are, generally, based upon the notion of "freedom of religion" which means a person has the freedom to choose what religion he will practice if he practices any at all. Thus, the existence and practice of religions are considered to represent an expression of a person's personal freedom. This status has been secured by giving religious organizations special constitutional protection and even tax free status. Fortunately, western political systems also provide the individual with "freedom *from* religion" by excluding their influence from secular government. This was done because the founding fathers in the USA had seen what could happen when particular religions managed to get a strangle hold on European governments over the centuries. The result was suppression of other religions, inquisitions, witch hunts, wars, and helping the wealthy to keep the poor "in their place" which usually meant in subservience to the upper classes.

But, what about the "good" that religions have done over the centuries?

Yes, they have done some good and it should not be completely ignored. For example, during the plague that struck Europe in the 14th century, Christian nuns would stay behind and tend to those who had become infected with the "Black Death", a terrible infection spread by the bites from the lice carried by rats that would ravage a person's body with fever and cover him with dark bruise-like blotches. This infection had a very high mortality rate and many of the religious that stayed behind to help the sick and dying while the still uninfected fled the cities were themselves killed by it.

And, yes, the belief in an afterlife where all of a person's troubles and worries would be a thing of the past can be very comforting to someone who is lying on their dead bed. It can also be comforting to those who are about to lose a loved one to think that they are not really dead and gone forever, but, rather, just "passing on" to some spiritual realm and they will eventually see him again at some time in the future when they also pass on.

While all of these good things religions have done certainly seem wonderful, one must remember that they all have a sort of selfish motive behind them. Those doing the good deeds and charitable acts do them because they expect to get something good in exchange. They imagine that there is a God up there in heaven who is watching them perform such acts and that it is all being recorded so that, when they die, it will help secure their own salvation. Thus, the good deeds are a form of salvation insurance! I've often thought that if an atheist did the same acts of kindness strictly for the sake of helping his fellow human without any expectation of (or believe in) being rewarded in some eternal and blissful afterlife, then his efforts would seem far nobler to any judgmental God than those of someone seeking a heavenly reward for such acts.

The clergy of many religions often claim that they are critically needed to help people worship a God who created the universe and without who's continuing favor and protection all manner of catastrophe and evil would be let loose upon Earth. Of course, this immediately brings one to the matter of the existence of God. If God does not exist, then that alone would seem to eliminate the need for any sort of religious clergymen.

Does God exist? That is a question that has occupied many thinkers over the millennia and it is not an easy one to answer definitively.

In ancient times the obvious answer would have been "yes" and, even today, this is still the "obvious" answer for most of the world's major religions. But, modern science says that it can explain the origin of our universe and the life it contains *without* the need to assume the existence of some supreme being who, one day, decided he, she, or it would make a universe. Those that demand that a god exist at "the beginning" to rationalize what we now observe always seem to avoid answering the simple question that a supreme god's existence demands: who created that god? They avoid having to provide an answer by just saying that God is "eternal" and has neither beginning nor end. The scientists, however, can then just as easily reply that the universe must also have that property.

As I suggested in one of my UFO books, *The How and Why of UFOs*, it's quite possible that our present "Big Bang" universe is just one of an infinite number of such expanding universes that each exists in the *same* infinite cosmos and extends out through space in all directions. Each of these coexisting universes begins its life at about the same time with the massive explosion of a giant, solar system sized Super Black Hole which can be referred to as a "Cosmic Egg" and whose ejected energy and matter eventually forms all of the galaxies, stars, planets, and life contained in that particular universe (the ability to perform the require release of this energy and matter is accomplished by a hypothetical "black hole escape mechanism" that I provide in the book).

Over the course of tens of billions to, perhaps, a hundred billions of years or more, the hundreds of billions of stars of each of the outward hurtling galaxies of each universe finally die out as their nuclear fuels are completely exhausted. Then, with the passage of another few tens of billions of years, the outer, roughly spherical boundaries of each still expanding, but dead universe begin to collide and overlap with those expanding outer boundaries of its neighboring universes. This then causes the dead stars of their colliding and intermingling galaxies to coalesce under the action of mutual gravitation and form hundreds of billions of black holes. These black holes then also begin to combine to form even larger black holes and, eventually, the entire universe disappears into an infinite collection of super black holes that drink in and concentrate any remaining electromagnetic radiation traveling through space. These super black holes will roughly be arranged into a regular three dimensional array throughout the cosmos which I've always imagined to be like the arrangement of atomic nuclei inside of a cubic crystal lattice.

Apparently, as the super black holes in this infinite array continue to pull additional matter into their event horizons and thereby continue to increase in mass, a time is reached when they all become structurally unstable at about the same time. At that time, this infinite number of super black holes then becomes an infinite array of Cosmic Eggs and, as they all begin to "hatch" and violently eject their matter and energy, an entirely new cosmos composed of an infinite number of fresh expanding Big Bang universes is created.

The entire cyclic process that I've just described is automatic and perpetual. It had no beginning or end in time or space and it will have no end. And, even more importantly, it is the *only* possible structure that our cosmos can have (which is one of the reasons that I do not believe in "higher" and "lower" dimensional "planes" that somehow invisibly coexist with our four dimensional space time continuum). This is a very important point that I feel the need to emphasize. Our cosmos *must* have this structure because it is *not* possible for it to have any other. If we try to imagine a cosmos that is devoid of all matter and energy, then that cosmos would also immediately be devoid of all space and time because these can not exist without the presence of matter and energy! Thus, an "empty" cosmos containing no matter or energy is actually a physical impossibility. It would immediately annihilate itself by shrinking to a point with infinitesimal volume and infinitesimal existence. It would just blink out of existence as our present infinite cosmos rushed into existence to take its place.

These, indeed, are weighty metaphysical issues to ponder.

With these basic cosmological concepts in mind, we see that there is really no need for a supreme being to create and maintain our universe. It does that itself automatically, perpetually, and mindlessly.

The theologians are also fond of telling us that God represents "infinite" love and concern for the welfare of "good" humans so that their invisible souls can live blissfully on after physical death in some idyllic afterlife or heaven. When one reminds them of all of the truly evil things that have been done to innocent people in this world over the millennia where this imagined all loving, all powerful god did nothing to intervene, they will usually say that this evil was all inspired by God's arch enemy, a fallen angel and now the leader of hell's demons named Satan, and that if the human evil-doers had been able to resist his influence because they had been more devout, then all would have been well. Some of the theologians will even suggest that God *purposely* lets evil take place on Earth because it serves some "higher" purpose such as preserving human beings' freedom of choice to do either good or evil. In any event, they claim, any innocent victims of such human evil will surely be greatly rewarded in heaven by God for the suffering they might have endured on Earth. I suspect that this type of "reasoning" is not really that comforting to the parents of a small child who was killed in a senseless accident or who died from some rare disease.

It is an easy matter to dismiss the notion of the existence of a supreme deity by just stating that if he truly existed and wanted humans to follow some approved moral code of behavior, then this god could immediately make himself visible to *everyone* on Earth and communicate his desire to them. For example, he could cause the face of the moon permanently turned toward Earth to take on his appearance and he could then cause his moral principles to be carved into the side of every mountain on our planet. With feats such as this there would be no doubt in anybody's mind that they were in the presence of some truly supreme being. In addition, he could change the laws of nature so that whenever someone broke one of his laws, he would immediately receive some punishment. Perhaps he would feel some pain or would suddenly age by a certain amount of time per transgression. Habitual offenders would age rapidly and eventually experience permanent death and decay. Those that obeyed the moral code directly given to them by God would not age or experience any sort of illness and, as a result, would be eternally youthful and happy. If these "good" humans were killed in an accident, they would immediately be resurrected back to their normal selves.

But, sadly, we do not see these simple (for a god, that is!) solutions to providing humans with the proof of the existence of a supreme being who is in charge of the universe and its laws. Instead, we see a collection of self appointed humans who each decide how their fellow humans should and should not behave. There are no large scale, difficult to hoax miracles, but, rather, only small local incidents most of which can be faked or easily dismissed as natural phenomenon or simple distortions of far less impressive events caused by the telling and retelling of past incidents to each new generation of "believers". Ultimately, when one gets involved in any religion all he will see are humans, buildings constructed by humans, clerical garb designed and manufactured by humans, "holy" books printed by humans, and songs composed and sung by humans. One will never see any gods, angels, or demons anywhere.

I have been asked once or twice what would happen if everyone on Earth decided tomorrow to get rid of all religions? What would we replace them with? How would charitable acts be motivated and done? How would the sick and dying be comforted along with their survivors?

After giving the matter some thought, I realized that all religions could be replaced with a single, worldwide system that would be based on ethical, rather than moral principles. It would be based more on a philosophy that tries to induce people to do things that are in the best interest of *both* themselves *and* their fellow human beings. It would get rid of all of the false prophesies that only serve to frighten people and replace them with an awe and reverence for the universe and the life it contains.

Since the church facilities of different religions provide a sort of hub for a community's various activities such as social gatherings, weddings, funerals, etc., any system that replaces religion should also continue with these beneficial services although they would be ones that are more in agreement with the observable and verifiable characteristics of our world and the cosmos we inhabit.

For example, weekly worship services to a supreme deity might be replaced by weekly gathering in which a secular clergy specially trained in matters of ethics, psychology, and science could give a short sermon that would instruct the masses on how to best behave so as to be as beneficial to all life on Earth as well as in our galaxy. Perhaps an hour long science documentary could be shown that would provide information about the formation of our Big Bang universe, the galaxies it contains, and the star systems within our Milky Way galaxy. People would be encouraged to think of themselves as well as all other life forms as a small, but rather unique *part* of the universe and not somehow separate from it. Humans, animals, and extraterrestrials would all be seen to be composed of atoms whose nuclei had originally been created by fusion reactions within their particular solar system's sun. Most likely, we will find that all animal life throughout our universe and those surrounding us is the product of adaptive mutations to chromosomes made from the same type of DNA molecule. The ultimate unity of all life, even plant life, will be emphasized and always considered to be part of the universe.

People with various physical, psychological, and social problems will be given free counseling at these weekly services or any time of the week if their need was great enough. They could then be given either immediate short term financial help if that was needed and then referred to others who might be able to provide them with more information about how to relieve their problems.

I firmly believe that as we develop the technology demonstrated by the UFOs now visiting our planet, we will, eventually, be able to treat all medical problems nearly instantly and even, as incredible as it may sound, resurrect the dead to eternal and youthful life. Until that time comes, however, people who were about to die from disease or old age could be assured that their remains would be carefully preserved by dehydration and stored at their local church's storage facility. They would know that, as soon as technology allowed it, they would be resurrected so that they could be with their loved ones again on an Earth finally freed, through UFO technology, from the various problems that currently plague us. They will no longer fear death, but look at it as only a temporary inconvenience. They will trust Earth's advancing science and technology, not an imaginary supreme being, to assure them a pleasant eternity right here on a planet and not in some vague spiritual realm.

Is what I have described here just a nice, but unattainable dream?

Well, if UFOs are real, then what I understand of their technology means that this new global secular "religion" based solely on science and ethics *is* a possibility. What will guarantee its success is if, as I predict, UFO technology allows for the resurrection of dead, even long death, human beings. Once that single process is shown to be possible, then all humanity will finally be freed from the present day collection of myths and superstitious delusions that have served to impede human progress for millennia now while subtly encouraging many of the wars that have ended millions upon millions of lives. If this had not happened, then one wonders what heights humanity might have reached by now. Who knows, perhaps it would be human beings who would be operating incredibly swift disc shaped craft in the skies of *other* worlds while their sentient inhabitants looked on in awe!

Chapter 11

Achieving Instantaneous
Interstellar Communication

I'M ALWAYS FASCINATED WITH HOW communication through vast distances of space is handed in old science fiction movies.

Usually, it is treated as though it is instantaneous regardless of the distances involved. For short distances, as from the Earth to our Moon, the delay of a second or so is virtually unnoticeable and can, for the purposes of movie making, be made instantaneous or without any delay. But, then we have movies in which flight controllers on Earth or in a space station orbiting Earth are attempting to communicate, via audio frequency modulated radio waves, with colonists on Mars or astronauts aboard a spaceship on the other side of our solar system. These communications, using radio waves which, like visible light, only travel at about 186,000 miles per second, should take hours to reach such destinations, but are also usually depicted as being instantaneous.

The only film in which I've ever actually seen some scientific attempt made to overcome this communication delay problem was a classic, big budget science fiction film from the 1950's titled "Forbidden Planet". This film is unique because it was the first to show human beings exploring a planet located in *another* solar system using a saucer shaped craft of their own construction.

The film's astronauts soon realized that they had to rescue a scientist and his daughter from a "planetary force" that was an incredibly strong and invisible being who did not like strangers on its planet and would not hesitate to tear them to shreds if it got a chance. Because of this situation, the captain of the Earth spacecraft needed to contact Earth for some instructions as to how he should proceed, but he was about 17 *light-years* away from the Earth on a planet orbiting the star Altair and, obviously, could not wait a total of 34 years for a response to a message when everyone's lives were in danger. The captain, however, had a chief engineer aboard the ship who mentioned using the ship's parts to build a powerful transmitter that could "short-circuit the continuum on a five or six parsec level" and thereby allow them to communicate with Earth in a timelier manner.

I won't discuss what happened in this interesting film for the sake of anyone who has not seen it yet, except to say that the transmitter was not completed and communication with Earth was not established.

Yet, I found the idea of near instantaneous or even instantaneous communication across the vast gulfs of space that separate even the closest of star systems within our Milky Way galaxy to be an intriguing one. During my decades of UFO research, I wondered just how our extraterrestrial visitors handled their ship to ship and ship to home base (whether it be in our solar system or another) communication.

For short range communication of a few hundreds of thousands of miles, they would, like humans, rely upon such things as varying the intensity of radio waves using audio frequencies to carry speech. This technique can also be used to vary the intensity of laser beams which, when in the visible portion of the electromagnetic spectrum, can be easily visually directed so as to strike a distant receiver's optical sensor. The biggest drawback with optical laser beams as opposed to radio waves is that the laser beams can be easily stopped by solid obstacles such as mountain ranges or buildings. They can also be refracted when passing through layers of atmosphere of different temperatures and densities which can then make accurately directing them difficult.

Another very advanced method of short range communication, perhaps only used for distances of less than a hundred miles, involves varying the intensity of the *excess* anti-mass field radiation that is emitted from the hull of a massless, airborne UFO. This new form of non-electromagnetic radiation, like electromagnetic radio waves, can also be varied at audio frequencies to carry speech and is probably responsible for the handful of reliable cases in which people reported hearing sounds being produced by the speakers of electronic equipment that was actually turned off!

While all of these methods are suitable for use as the crew of a mother ship exploring a solar system's planets tries to stay in near instantaneous communication with the smaller craft they've dispatched into a particular planet's atmosphere, they all have undesirably long delay times when one tries to use them to communicate between the inner and outer planets of a solar system or, obviously, between two widely separated solar systems.

One possible solution to this problem would be for an alien mother ship to simply record a message into the computer of a small "messenger probe" and then have it fly at enormous hyper-light velocity to the intended recipients of the message in another part of a solar system or even in another solar system entirely. Once the probe had delivered its message to them, the recipients would, in similar fashion, upload their reply and the probe would then immediately return at hyper-light velocity back to the craft of its origin.

While this system is workable, it is certainly not really instantaneous. Even if such a probe could achieve a hyper-light velocity of a million times the velocity of light, it would still take it about 2.1 minutes to travel a distance of 4 light years which is about the average distance between solar systems in our galaxy. If an extraterrestrial mother ship's home system base was, say, 100 light years from our planet, then it would take the probe 52.6 minutes to deliver each message. This, of course, assumes that nothing happens to the probe as it streaks toward its destination and that it experiences no delay in finding the location of a recipient and downloading its message.

But, what I was looking for was a simple device that would allow truly instantaneous communication between the separate solar systems within a galaxy or between separate solar systems in *different* galaxies. Could such a thing be possible?

Incredibly, the answer is "yes"!

In the early days of the development of quantum mechanics, the branch of physics that describes the processes that occur on the microscopic level of atoms and subatomic particles, an effect called "quantum entanglement" was discovered. The effect was predicted theoretically and only years later verified experimentally. When Einstein found out about this effect, he immediately took a dim view of it because it appeared to violate his famous Theory of Relativity. That theory states that neither matter nor information can be transmitted through space in excess of the speed of light. In fact, Einstein, together with fellow physicists Podolsky and Rosen, published a scientific paper in 1935 in which they tried to prove that quantum entanglement did not make sense and was impossible. They could do this because, at the time, there was no experimental proof that the effect was real. Einstein even derided this important quantum mechanical effect by referring to it as "spooky action at a distance". In general, Einstein was a major opponent of quantum theory because it seemed too random and capricious to him so it is little wonder that he would have adopted a negative opinion of quantum entanglement.

Basically, quantum entanglement is the effect noted for such microscopic entities as electrons, molecules, nuclei, and photons whereby certain of their physical properties such as momentum, polarization, position, and spin can become "correlated" or, as the effect's name implies, "entangled". This means that if anything happens to pairs of these particles after they are formed that causes one member of the pair to undergo a spontaneous and sudden change in its state, then the other member of the pair will *instantly* undergo a similar change in the *opposite* direction so that the two changes cancel each other out. More importantly for the purposes of this chapter, the change that takes place in the other member of the pair will occur *instantly* no matter how great the physical distance between the two particles! Thus, if the two particles happen to be separated by *any* amount of distance, one will always have changes taking place in the other member of the pair which will *infinitely* exceed the velocity of any sort of information that could travel between the two at the velocity of light. As mentioned above, this seems to violate one of the foundational principles of Relativity Theory which states that nothing can travel faster than light, not even information.

In thinking the matter over, however, it occurred to me that it should be a somewhat easy matter to create an instantaneous interstellar communication device based upon the quantum entanglement effect. To construct such a device, one would need to split a large quantity of helium atoms into two equal quantities of hydrogen atoms. Each of these resulting two supplies of hydrogen atoms would be held in a separate container and, regardless of the distance placed between them, certain changes made to the nuclei of the hydrogen atoms in one of the containers would instantly result in the opposite effects taking place in the nuclei of the hydrogen atoms in the other remotely located container. If that happens, then is would be a simple matter to have the changes taking place in the hydrogen nuclei of the first container vary in response to an information carrying audio or digital signal and to then have the exact opposite of those changes *instantly* taking place in the hydrogen nuclei of the second container which could actually be located in a neighboring galaxy!

Before I give the details of the design for such a device, however, the reader needs to learn a few things about the structure of atomic nuclei, particularly helium and hydrogen nuclei.

For example, imagine that we took a single helium atom and then, by bombarding its nucleus with enough energy in the form of electromagnetic radiation, caused it to split in half to form two hydrogen nuclei.

The original helium nucleus contained two positively electrically charged protons and two electrically neutral neutrons. (Each neutron can be thought of as a combination of a positively charged proton and a negatively charged electron which explains why it is electrically neutral. Perhaps the proton and electron that make up a neutron are in actual physical contact with each other or the electron, being less massive, orbits the much more massive proton of the neutron at an enormous velocity.) After being split in half, the two hydrogen nuclei produced will each contain a single proton and a single neutron.

Note that the negatively charged electron in each neutron serves to pull two protons together inside of an atomic nucleus so that the powerful repulsive electrostatic forces caused by their extreme proximity to each other will not make them fly apart and thereby cause the nucleus to undergo fission. Neutrons are critically important to the stability of nuclei such that whenever one has more than one proton present in a nucleus, he *must* also have one or more neutrons present in that nucleus if it is to be stable.

If there is an insufficient supply of neutrons present for the number of protons in a particular atom's nucleus, then the nucleus will be unstable and can eject some of its protons in the form of helium nuclei called "alpha particles". This ejection of protons then increases the ratio of neutrons to protons in the nucleus and helps to make it more stable. Of course, the loss of any protons from an atomic nucleus immediately causes the nucleus to become that of a different and less massive element. This is referred to as a "transmutation" of the original atom.

For a much more massive nucleus then that of the hydrogen atom, it is possible for a lowest level electron orbiting the nucleus to occasionally be "captured" by a proton in the nucleus such that the electron then joins the capturing proton to form a neutron. Any protons left over inside of a nucleus which are *not* found inside of the electron / proton combinations which make up the nucleus' neutrons can then be referred to as "non neutron protons". This process also results in the nucleus emitting some gamma radiation as the capturing takes place. In this case, nuclear stability is achieved by *increasing* the number of neutrons inside of an atom's nucleus while simultaneously *decreasing* the effective number of protons as those protons become bound to electrons.

All protons as well as electrons can be imagined as physically spinning and this motion will give each proton its own little magnetic field since a magnetic field is produced whenever an electrically charged particle moves through space or rotates about one of its internal axes. It's simple to visualize the relationship between a *positively* electrically charged spinning proton and the magnetic field that its motion produces. If one curls the fingers of his *right* hand against its palm and extends its thumb, then the direction his fingers curl would represent the direction in which the equator of a spherical proton spins and the direction his thumb points in would then represent the direction that the magnetic field lines emerge from the proton's north magnetic pole. The magnetic field lines used to graphically represent that tiny magnetic field would then curve around the outer spherical surface of the proton until they converged and reentered the proton at its south magnetic pole. This south magnetic pole would be located diametrically opposite from the proton's north magnetic pole.

When dealing with *negatively* electrically charged subatomic particles like electrons that are spinning, one must use a "left hand rule" so that if the curled fingers of one's *left* hand represent the direction that the spherical body of an electron spins, then that hand's extended thumb would represent the magnetic field lines as they emerge from the north magnetic pole of the electron. Note that when an electron and proton are spinning in the same direction (i.e., both clockwise or both counter clockwise) and are in close proximity to each other as happens when they form neutrons, their external magnetic fields will point in opposite directions. This causes them to become magnetically coupled to each other so that their individual magnetic fields do not extend that far out into space. This is why individual neutrons have little, if any external magnetic fields and also have no *net* spin.

Thus, each submicroscopic spherical proton in the helium nucleus can be considered to have a spot on its outer surface from which its magnetic field emerges. That magnetic field then diverges in all directions and loops around all of the sides of the spherical proton. When the magnetic field reaches the exact opposite side of the proton, its magnetic field lines then converge and dive back into a spot there on its surface. From this we see that each spinning proton is actually similar to a tiny bar magnet that has a north and south pole with its north pole being the spot on its surface from which the particle's magnetic field lines emerge and its south pole being the spot on its surface into which the external magnetic field lines return.

Currently, it is possible to split helium nuclei into separate hydrogen nuclei and thereby create two hydrogen atoms from each helium atom. However, the process requires bombarding the helium nuclei with huge amounts of energy and, unfortunately, our current technology only allows this process to form microscopic quantities of hydrogen nuclei each of which then quickly captures a single electron to form an electrically neutral hydrogen atom. After such artificially produced hydrogen atoms are formed, they can, upon collision, further bond to each other so as to form gaseous hydrogen molecules. Each such electrically neutral molecule will contain only two hydrogen atoms and is referred to in chemistry as a "diatomic" molecule.

Before one can construct an effective instantaneous interstellar communication device, however, one must be able to split huge numbers of helium nuclei into twice as large a number of hydrogen nuclei. Ordinarily, the amount of energy that would be required for this would be on a par with the amounts of energy released when hydrogen bombs are detonated since one has to literally supply the helium nuclei with the same enormous amounts of energy that are released when hydrogen nuclei are fused together to form helium nuclei during a thermonuclear explosion or when this fusion happens inside of a star! Obviously, such a process would make creating the required amounts of hydrogen nuclei impossible in a laboratory or factory environment.

Why not just use the nuclei from regular and easily obtained hydrogen gas to make the instantaneous communication device?

The reason is that we must have two large, *separate* samples of hydrogen gas such that the single non neutron proton in *each* hydrogen atom nucleus in one of the samples is correlated with or in a state of quantum entanglement with another single non neutron proton of a hydrogen atom nucleus in the second sample from which it was parted by the process of fission. Unless this can be done, then it won't be possible to construct such a communication device. Fortunately, there may be a way to obtain these large samples of entangled nuclei that does not require the expenditure of huge amounts of energy. Unfortunately, however, this method will not be feasible until humanity has mastered the construction of the anti-mass field generators that make the inertialess, massless flight of UFOs possible. Despite this limitation, however, it is still possible for us to theoretically, at least, explore this novel method for obtaining huge supplies of quantum entangled hydrogen atom nuclei.

In earlier chapters I mentioned how the anti-mass field radiation that is emitted from an airborne, massless UFO's anti-mass field generator comes in several types or "characters" each of which determine various unique physical effects for this new form of nonelectromagnetic radiation.

There was the most prevalent or "neutral" character type found in the anti-mass field radiation used to greatly reduce or even completely extinguish the mass and, thus, the inertial and gravitational properties associated with a UFO and its crew. This character of anti-mass field radiation was essential to allowing UFOs to perform the dazzling aerobatic maneuvers they often display in Earth's atmosphere or, when traveling through outer space, to readily achieve hyper-light velocities far in excess of that of light. Then, in addition to anti-mass field radiation with neutral character, there were also artificially produced anti-mass field radiations with "B rich" and "E rich" character.

The B rich anti-mass field radiation had the remarkable ability to reduce the electrostatic force of *repulsion* that acted between two subatomic particles with the *same* electrical charge polarity such as the force that exists between either two negatively electrically charged electrons or between two positively electrically charged protons. B rich anti-mass field radiation could be used to allow one object to actually physically interpenetrate and then pass completely through another object.

As we saw earlier in this book, B rich anti-mass fields could be biologically generated by some of Earth's subterrestrial inhabitants and allowed them to pass through the solid rock walls that separated subterranean cavities. They could also be used by them to interpenetrate their way through closed doors and walls so that they could easily enter and explore human surface dwellings. There is evidence in the UFO reports that this effect is also used by two or more UFOs that physically overlap portions of their hulls while in flight so as to form a single composite craft. There are even a few cases that describe UFOs that physically interpenetrate the ground and disappear into it! Most likely, these cases are instances of craft that are passing through the Earth's crust to the safety of some deep subterranean base that they have established.

One of the most important uses of B rich anti-mass field radiation is when it is used to surround a massless, airborne UFO with a layer of highly ionized atmosphere or plasma. The various ions in this plasma can then be made to flow rapidly around an airborne massless craft's outer hull by projecting crossed or perpendicular electric and magnetic fields into the layer containing the ions. These fields, when crossed, will then apply Lorentz forces to the ions in the plasma to make them move around the outside of the UFO's hull. In this case, the UFO's anti-mass field generators do not directly make the B rich anti-mass field radiation. Rather, it is automatically formed whenever excess neutral anti-mass field radiation leaves the surfaces of the craft's outer hull and passes through any of the projected magnetic fields near the hull.

This B rich anti-mass field radiation, however, does not behave like the usual kind that is *directly* produced inside of an anti-mass field generator. This is because it actually induces ionization in the atmospheric layer surrounding the craft's hull by reducing the electrostatic force of *attraction* that exists between subatomic particles with *opposite* electrical charges such as exist between negatively charged electrons and positively charged protons instead of, as happens with regular B rich anti-mass field radiation, reducing the electrostatic force of *repulsion* that exists between subatomic particles with the *same* electrical charge such as exists between two electrons or between two protons.

I consider this a paradoxical effect and have dealt with it in my book "The How and Why of UFOs". Suffice it to say here that this unusual form of B rich anti-mass field radiation, which is only created as a secondary effect outside of an airborne massless UFO's hull, is very important to the functioning of the craft's plasmadynamic propulsion system. Its existence is also important to understanding how UFOs can cause electrical equipment to malfunction by saturating the region with them with plasma that then allows unexpected high voltage short circuits to occur. In these cases, this unusual form of B rich anti-mass field radiation behaves exactly as would E rich anti-mass field radiation that had originally been produced inside of an active anti-mass field generator. Hopefully, the reader will not be confused by the differences between regularly produced and unusually produced B rich anti-mass field radiation. We need not consider either type further in this chapter.

This brings us finally to the E rich anti-mass field radiation directly produced inside of anti-mass field generators which always tends to reduce the electrostatic force of attraction that exists between two subatomic particles of opposite electrical charge such as that which exists between a negatively charged electron and a positively charged proton.

As was seen in an earlier chapter, it is E rich anti-mass field radiation that allows both UFOs and living beings to achieve a state of optical invisibility which simply means that, although physically present, the various atoms of their structures no longer can interact with photons of visible light so as to make these structures visible to other beings who are using visible light for their vision. E rich anti-mass field radiation creates this awesome effect by weakening the electrostatic force of attraction that exists between a positively charged atomic nucleus and the various electrons which orbit it at high velocity. Weakening the attraction of an electron for its nucleus allows the electron to move a little farther from its nucleus and into a new orbit with a slightly larger radius. Since all of the electrons in an atom exposed to an E rich anti-mass field are simultaneously affected, they all tend to move a bit farther from their common nucleus.

This expansion, however, causes a change in the energy *difference* between any two orbits within any atom such that this difference is always reduced a little. Because of this reduction in the amount of energy needed to move electrons from lower to higher orbits within an atom, the electrons being affected by an E rich anti-mass field will only absorb photons of electromagnetic energy with lower frequencies and thus energies when they need the energy from them to make such an orbital transition within the "cloud" of electrons that surrounds an atom's nucleus. At some point, when the E rich anti-mass field radiation passing through an object is intense enough, any electronic transitions that take place inside of the atoms of the object will be performed completely by the absorption of photons of electromagnetic radiation that are lower in frequency and energy than those found in the visible portion of the spectrum. At this point the object, whether inanimate or living, will become completely invisible to an eye using visible light for vision. However, it is certainly possible for the object to be seen by an instrument or being who can "see" in the infrared band of the electromagnetic spectrum.

It may be possible to use E rich anti-mass field radiation to fairly easily split large quantities of helium nuclei into their component hydrogen nuclei and, more importantly, keep the two non neutron protons in each such pair of "daughter" nuclei formed "spin correlated" as this is done which just means that the two hydrogen nuclear non neutron protons in each split apart pair will remain quantum entangled no matter how far the two nuclei are physically separated from each other. Here's how this might be done.

Imagine that a tank of ordinary helium gas is connected via a hose to an evacuated chamber which is located near a small anti-mass field generator. The anti-mass field generator has its operational parameters adjusted so that the chamber can be penetrated by intense E rich anti-mass field radiation (exactly how E rich anti-mass field radiation is created inside of an anti-mass field generator is discussed in detail in my book *The New Science of the UFO* and involves adjusting the ratio between the intensities of the electric and magnetic fields inside of the active device which are used to produce its antigraviton emission).

When the valve on the tank is opened, individual helium atoms will begin to enter the evacuated chamber and can actually be made to spray into it from a nozzle located in the side of the chamber. Since the chamber is also penetrated by intense E rich anti-mass field radiation, we can expect the electrostatic forces of attraction acting between *each* helium nucleus and its two orbiting electrons be greatly weakened and these electrons will begin to orbit at a slightly farther distance from their nucleus. However, there should also be an effect taking place *inside* of the nucleus of each helium atom inside of the chamber.

Each helium nucleus contains two protons (which, as previously stated, can be referred to as non neutron protons) and two neutrons, but, since each neutron can be considered to be a proton paired up with an electron, we can actually consider the helium nucleus to be made up of four protons and two electrons. The protons are all positively electrically charged and exerting enormous repulsive forces on each other due to their proximity within the tiny nucleus of the atom. However, they don't fly apart because there are two electrons that are located in their midst. These electrons, being negatively electrically charged, exert equally powerful attractive forces on the positively charged protons and manage to keep all six particles in a stable configuration with respect to each other.

This configuration is so stable that, ordinarily, one would have to put a lot of energy into it in order to allow the nuclear particles, sometimes referred to as nucleons, to begin to separate and fly out of the nucleus. Such energies are achievable inside of devices developed in the 20th century known as particle accelerators. They use high voltages to accelerate charged particles like protons to very high velocities so that they have very high kinetic energies. If one of these high speed protons is made to collide with a helium nucleus, then fission can occur and the helium nucleus will be split into hydrogen nuclei. Unfortunately, the process can only produce minute quantities of hydrogen nuclei.

Using E rich anti-mass fields may, however, dramatically lower the amount of energy needed to split helium nuclei. This is because, if the E rich anti-mass field is intense enough, then the stabilizing electrostatic forces of *attraction* acting between the two electrons and four protons inside of the helium nucleus can be weakened enough so that, perhaps, only the energy carried by photons of x-radiation need be added to a helium nucleus' protons in order to enable them to overcome the weakened attraction of the electrons that keep them inside of the nucleus. Thus, once helium atoms being penetrated by E rich anti-mass field radiation are also exposed to intense x-radiation, the helium nuclei inside of the atoms will begin to rapidly disintegrate and will form a collection of hydrogen nuclei. Each of these hydrogen nuclei will then pick up one electron from the previously existing helium atom and become an electrically neutral hydrogen atom and, when two of these two newly created hydrogen atoms come into contact inside of the chamber containing them, they will immediately form a single diatomic hydrogen *molecule*.

All of the hydrogen nuclei thus formed will be identical and each will contain one non neutron proton and one neutron. Since the neutron can be considered to consist of an electron bound to a proton, this means that each hydrogen nucleus actually contains two protons that are held together inside of the nucleus by a single electron. However, the newly formed hydrogen atoms will not exactly be identical to "normal" hydrogen atoms. They will, in fact, be slightly less massive than regular hydrogen atoms. This is because by using an E rich anti-mass field and x-radiation to produce the fission of helium nuclei, we did not have to input the normally required massive amount of energy to do so which would result in the production of hydrogen nuclei with their normal masses. In a sense, we cheated nature in order to conserve the amount of energy that had to be put into the helium nuclei during the fission process and thereby greatly increase the quantity of helium nuclei that could be split and the quantity of hydrogen nuclei that this would produce. The price we must pay for this cheating is the production of hydrogen nuclei that are a bit mass deficient as compare to normal hydrogen. (Note that the mass of the hydrogen nuclei is determined by how much energy was used to form them during the fission of helium nuclei. The less energy used, the more mass *deficient* the hydrogen nuclei will be. Consequently, any electrically neutral atoms or diatomic molecules formed from these less massive nuclei will also be mass deficient.) But, aside from this distinction, there will be no other differences in the chemical, electrical, magnetic, or physical properties of the hydrogen nuclei, their atoms, and finally the diatomic molecules that are produced.

As the stream of helium atoms enters the chamber and is simultaneously affected by the E rich anti-mass field radiation and x-radiation penetrating that chamber, the helium nuclei will begin splitting in huge numbers. Depending upon the orientations of the E rich anti-mass field radiation and x-ray beam entering the chamber, this stream of helium nuclei can be made to split into two separate streams of hydrogen nuclei that will rapidly separate from each other.

The artificially induced fission process each helium nucleus experiences will cause the non neutron protons of the two hydrogen nuclei formed to spin in *opposite* directions and this is required by Newton's third law of motion which is sometimes referred to as the "Conservation of Momentum Law". Because each hydrogen nucleus contains a positively charged non neutron proton in it, this proton's spinning about its internal axis will cause each hydrogen nucleus formed to develop a tiny magnetic field (the proton contained in a hydrogen nucleus' single neutron also has a magnetic field, but it is cancelled out by the opposing magnetic field of the spinning electron in the neutron).

Since the two non neutron protons in the two hydrogen nuclei formed are spinning in opposite directions, their two spin axes and their resulting magnetic fields will not be aligned, but, rather, will be pointing in opposite directions. That is, we can imagine that if the north pole of one hydrogen nucleus' non neutron proton is pointing up, then the north pole of the other hydrogen nucleus' non neutron proton's north pole will be pointing down. Despite the presence of these coupled nuclear magnetic fields, however, the two hydrogen nuclei formed will not be pulled back together again by the attractive magnetic force acting between the separating nuclei because it is simply not strong enough to do this although this force may slow down the speed at which the nuclei separate from each other. At this point, the spin directions of the two non neutron protons in each pair of separating hydrogen nuclei formed will be in opposition because their angular momenta *must* cancel each other out as required by the Conservation of Momentum Law previously mentioned. However, despite this, they will remain in a state of quantum entanglement with each other. The directions that their individual spin axes point in will also be in opposition.

As the two streams of hydrogen nuclei are formed and begin to separate, the two streams would *each* further be made to pass through a region penetrated by an external magnetic field provided by a permanent magnet.

The *two* external magnetic fields affecting the two parting streams of hydrogen nuclei would have the directions of their field lines oriented so as to maintain the *opposite* alignments of the hydrogen nuclei non neutron proton magnetic fields found in each of the parting streams. Thus, these two external magnetic fields would always have their magnetic field lines pointing in the *same* direction as the magnetic field lines exiting the north poles of the hydrogen nuclei non neutron protons in each of the two separating streams of hydrogen gas. These two required external magnetic fields, which can be provided by the new rare earth alloy magnets, are critically important to maintaining the quantum entanglement of the *opposed* spin directions of the non neutron protons contained in the two streams of newly formed, but mass deficient, hydrogen nuclei even as they each go on to almost immediately capture an electron to form a hydrogen atom two of which then go on to form a diatomic molecule.

To make portable instantaneous communication devices from the large amounts of quantum entangled hydrogen nuclei formed in our hypothetical fission chamber, it is necessary that the two newly formed sets of nuclei be separated from each other and then removed from the chamber and, most importantly, that the trillions of pairs of spin direction coordinated non neutron protons that exist in the two samples (with one member of each pair in each of the two samples) remain quantum entangled. Fortunately, this is relatively easy to achieve.

As the two streams of newly formed and quantum entangled hydrogen nuclei separate and go on to form electrically neutral hydrogen atoms and then diatomic molecules, they can each be made to flow into and collect inside of their own rigid, nonmetallic, tubular container which would be made from some kind of shatter resistant plastic material. Each of these two tubular containers will already have been equipped with two disc shaped, high strength rare earth alloy magnets at their opposite ends. These magnets will have their poles arranged so that they pass a strong *internal* magnetic field through the lengths of the tubular containers containing the hydrogen molecules. The two end magnets for each tube will, outside of the tube, be connected to each other by a thick, high magnetic permeability metal piece that is attached to their outer pole faces. This structure will act as an external flux bridge that contains the *external* magnetic field that exists between the two end magnets of each tube and helps assure that the internal magnetic field existing along the length of a tube is as strong and uniform as possible.

In essence, as *each* stream of newly formed gaseous hydrogen molecules collects inside of its own tube, the tiny nuclear magnetic fields of its trillions of non neutron protons will all transition from being in alignment with *one* of the two opposed external magnetic fields of the helium splitting chamber to being in alignment with the *internal* magnetic field inside of its particular tube which is provided by the permanent magnets attached to its ends. Obviously, just as the internal magnetic fields inside of the helium splitting chamber were in opposition to each other, the two magnetic fields provided to the two collections tubes by the end magnets attached to them will also initially be in opposition to each other.

If we call one of the tubes "tube A" and the other "tube B", then we can say that for every hydrogen nucleus in tube A with its nuclear magnetic field, due to the spin of the one *non* neutron proton in its nucleus, pointing in one direction along the length of its tube as it aligns with that tube's internal magnetic field, there will also be one hydrogen nucleus located somewhere in tube B that has its non neutron proton's magnetic field pointing in the *opposite* direction along the length of its tube as it aligns itself with that other tube's internal magnetic field.

Aligning all of the nuclear magnetic fields of the hydrogen non neutron protons in each tube in this manner also guarantees that the non neutron protons in the two tubes are constantly maintained in a state such that, *relative to the lengths of their tubes, all* of the non neutron protons of the hydrogen nuclei in tube A are always spinning in the opposite direction to that in which *all* of the non neutron protons in the hydrogen nuclei in tube B are spinning. And, of course, this also means that the direction that all of the spin axes of the non neutron protons in tube A point will always be opposite to the direction in which all of the spin axes of the non neutron protons in tube B point. Amazingly, this state will be maintained despite the chaotic motion of the trillions of diatomic hydrogen molecules in each tube as they collide with each other and rebound off of the cylindrical walls of their tube. More importantly, *every* single spinning non neutron proton of a hydrogen nucleus in tube A will always remain quantum entangled with *one* spinning non neutron proton in tube B.

When the tubes have been filled with a sufficient number of hydrogen molecules (all of which will be slightly mass deficient compared to normal hydrogen molecules), the tubes can then be sealed and removed from the helium nuclei splitting chamber. No matter how far apart the two sealed tubes are separated or what their eventual orientations are with respect to each other in space become, all of the spinning non neutron protons in one tube will always be spinning in the opposite direction to that of the spinning non neutron protons in the other tube. As long as quantum entanglement is maintained by not breaking the tubes of gas or removing the magnets attached to their ends, whatever affects the spin direction of the non neutron protons of the hydrogen nuclei in one of the tubes will also *immediately* cause the *opposite* effect to occur in the spin directions of the non neutron protons in the other tube. It is this effect, which takes place at infinite speed, which can allow for the construction of an instantaneous interstellar communication device.

Let's now see what else will be needed to complete the device once we have overcome the obstacle of producing, collecting, and stabilizing a large supply of spin direction quantum entangled hydrogen non neutron protons.

To build our instantaneous communication device, some means must be devised of causing some quantity of the non neutron protons in tube A to change their spin directions by changing the directions in which their axes of rotation point. That will then instantly cause their particular set of correlated non neutron protons in tube B to change the directions of their axes of rotation which will make them spin in the opposite direction to the non neutron protons in tube A.

Before the non neutron protons of the hydrogen nuclei in tube A are made to collectively and simultaneously change the direction of their spin axes, we can imagine that all of their spin axes are pointing up toward one end of the tube and that the magnetic field lines emerging from the trillions of north poles of these non neutron protons emerge from the spots on their *upper* spherical surfaces and then join the much more concentrated magnetic field lines that emanate from the north pole of the cylindrical rare earth magnet at the bottom end of tube A. All of these magnetic field lines then flow *up* toward and into the south pole of the other rare earth magnet at the top end of tube A. After they enter this magnet's south pole, they flow through this magnet, leave its north pole that faces away from the end of the tube, enter the high permeability flux bridge attached to this magnet, and then flow back down through the flux bridge *outside* of the tube, and finally reenter the south pole of the disc magnet attached to the bottom of the tube which also faces away from the end of the tube so as to form a complete magnetic circuit. At the same time, some of the magnetic field lines from the rare earth magnet attached to the bottom of tube A, as they emerge from its north pole which faces toward the tube and flow *upward* toward the top end of the tube, are diverted and enter into the south pole spots on the bottom surfaces of the trillions of non neutron protons inside of the tube.

In tube B the identical process is also taking place, but it is inverted with respect to tube A. Thus, in tube B the magnetic field lines exit the north pole spots on the *lower* spherical surfaces of the trillions of non neutron protons inside of its hydrogen nuclei, but then join the far more concentrated magnetic field that exists between the rare earth magnets attached to the ends of the tube and flow *down* toward the bottom of the tube. There they enter the south pole of the magnet at the bottom of the tube, leave its north pole which faces away from the tube, flow through this tube's external flux bridge, reenter the south pole of the rare earth disc magnet attached to the top end of the tube which faces away from the tube, and, finally, exit the north pole of this magnet which faces into the tube and begin flowing *down* along the length of the tube toward the disc magnet attached to its bottom end. Along the way some of these magnetic field lines are diverted and reenter the south poles spots on the upper spherical surfaces of the trillions of non neutron protons in the tube. Once again, we have a complete magnetic circuit.

Now, no matter how far the separation between tubes A and B or their orientations with respect to each other, it is a relatively simple matter to send instantaneous messages between them. Let's imagine that tube B had been put aboard a massless UFO and flown at an almost unimaginably high hyper-light velocity to the other side of our Milky Way galaxy so that it was separated by a distance of 100,000 light years from tube A which is located in an electronics laboratory on Earth.

Depending upon the strength of the two disc shaped, rare earth alloy permanent magnets' magnetic field inside of tube A which is used to orient the spin axes of its trillions of hydrogen nuclei non neutron protons by causing their tiny magnetic fields to all simultaneously align with that field, a certain amount of energy will be needed *per* non neutron proton in order to make it flip over and then have its miniscule magnetic field point in the exact opposite direction to that of the rare earth alloy magnets' magnetic field inside of the tube (that is, to make the north magnetic poles of a non neutron proton change from pointing toward the top of the tube by flipping or rotating it through an angle of 180 degrees so that it then points toward the bottom of tube A). If there was *no* state of quantum entanglement between the non neutron protons in tubes A and B, then this energy could easily be provided to *each* non neutron proton in tube A in the form of a single photon of electromagnetic energy which would have a certain low frequency in the radio wave region of the spectrum.

To provide photons of radio wave frequency electromagnetic energy to tube A only requires that it be wrapped in a metal coil which has a small alternating electrical current flowing through it with the appropriate frequency. If one of the energized coil's emitted radio frequency photons entered tube A and was fully absorbed by a non neutron proton in a particular hydrogen nucleus, then that proton would immediately flip over through an angle of 180 degrees because it would then have the energy needed for its tiny magnetic field to fully resist the aligning torque of the tube's internal magnetic field from the two rare earth alloy magnets as it flows along the length of the tube. The reorientation of the spin axis and its resulting magnetic field of this proton, however, will only be temporary. After a very short period of time, it will then spontaneously flip *back* through 180 degrees again so that its spin axis and the tiny magnetic field associated with it once again align with the more powerful internal magnetic field flowing along the length of the tube between the permanent magnets attached to the ends of the tube. If this proton does not receive another incoming radio frequency photon, then the directions of its spin axis and magnetic field will remain in this starting orientation.

As this restorative flip spontaneously occurs, the proton will release a radio wave photon of the exact same frequency as the one it originally absorbed that made it originally flip its spin direction that its tiny magnetic field opposed the internal magnetic field of the tube provided by its end magnets. That released photon will then continue moving in the same direction as was possessed by the original photon. Obviously, since the low pressure diatomic hydrogen gas molecules in tube A are each exposed to many radio wave photons of the correct frequency and energy, practically all of the trillions of non neutron protons in the tube will be flipped into the orientations in which their magnetic fields are opposed to that of the permanent magnets attached to the ends of the tube.

If the transmission coil is turned off suddenly, then all of the misaligned magnetic fields of the trillions of non neutron protons in tube A will suddenly begin "relaxing" or spontaneously flipping back into their original orientations that align them with the internal tube magnetic field of the rare earth permanent magnets mounted on the ends of the tube. As this relaxation occurs, the trillions of non neutron protons flipping back into their starting orientations with respect to the stronger alignment field inside of tube A will then actually act as a transmission antenna and emit their own burst of radio frequency photons that will exit the tube in all directions radially about its length. The coil that is wrapped around tube A can then act as a receiving antenna and absorb and convert these photons into a momentary alternating current that can be displayed on an oscilloscope screen as a brief AC voltage spike.

In the above analysis we have forgotten about the *other* tube, tube B, which is located on the other side of our galaxy! When that tube and its trillions of spin correlated hydrogen nuclei non neutron protons are considered, the analysis above must be modified.

When tube B and its trillions of spin correlated non neutron protons are considered, we would find that if we just bombarded the trillions of non neutron protons in tube A with the exact number of photons needed to make them all flip over so that their tiny magnetic fields all opposed their tube's internal alignment magnetic field provided by the tube's end magnets, then nothing would happen! This is because each of the non neutron protons in tube A is, via the quantum entanglement effect, for all practical purposes actually physically connected to a "sister" particle in tube B which is located on the other side of our galaxy. Since both particles must flip at the exact *same* time, this means that *each* non neutron proton in tube A must absorb a radio wave photon that carries enough energy to accomplish the flipping of *two* linked particles even though that linkage is invisible and still very mysterious in nature. Since the energy carried by a photon of electromagnetic energy is proportional to its frequency, this will require that each of the quantum entangled non neutron protons in tube A absorb a radio wave photon that has double the frequency of the ones that would be used if there was no quantum entanglement between the particles in tubes A and B or if tube B did not exist.

Imagine that we have adjusted for the spin correlation between the non neutron protons in tubes A and B and have, accordingly, doubled the frequency of the radio wave photons that the transmission coil surrounding tube A delivers to that tube's gaseous contents. When these higher frequency photons are emitted from the coil, they will simultaneously cause the non neutron protons in *both* tubes to flip over so that their tiny magnetic fields all point in the opposite direction to the alignment fields in the two tubes. Once that has happened, imagine that the coil surrounding tube A is disconnected from its source of alternating current and no longer sends any photons into tube A. What will happen when this is done?

Incredibly, as the orientations of the spinning non neutron protons of the hydrogen nuclei in the two vastly separated tubes spontaneously relax so that their spin axes and their resulting north pole magnetic fields again align with the internal magnetic fields provided by each tube's attached end magnets, *all* of the radio wave frequency photons that were absorbed by the nuclei in tube A will be readmitted by those in tube B! Thus, the tube A non neutron protons flip in their tube's internal magnetic field without releasing any radio frequency energy, while their sister particles in tube B account for all of the readmitted radio wave photons. This certainly seems odd, but apparently the cosmos allows it because this effect serves to cancel out the equally odd initial situation in which the particles in tube B were made to flip over against their tube's internal magnetic field without needing to directly absorb any photons. We see from this that the cosmos is satisfied as long as there is symmetry to events occurring within it even though they are not exactly local (but are, of course, when one considers the cosmos to be infinite in size!) and no energy is gained or lost as the process takes place.

With the quantum entanglement that exists between the non neutron protons in the two tubes, the reader may now realize how such a simple device could be used for instantaneous communication.

Imagine that a person on Earth decides to send a simple message using tube A to another person located on the other side of our galaxy that is near tube B. The person on Earth would merely have to energize the transmission coil surrounding tube A which would then cause the non neutron protons in both tubes A and B to simultaneously flip over so their spin axes and tiny magnetic fields all opposed their tube's internal alignment magnetic field. Then the person on Earth shuts off tube A's transmission coil. When this is done, the sister hydrogen nuclei in both tubes will nearly immediately relax back into their original orientations which align their tiny magnetic fields with their particular tube's internal alignment field. When this happens, there will be a burst of radio wave frequency photons emitted from tube B. That burst will then cause a brief pulse of alternating current in the coil surrounding tube B which can be detected on an oscilloscope and even turned into an audible sound.

If the operator of the tube A device on Earth attaches a switch to the coil that energizes the spin correlated non neutron protons in both of the tubes, then he cause pulses of radio frequency photons to be emitted from tube B whenever he chooses. This simple method can then be used to vary the time duration between any two sequential pulses that are emitted from tube B. One can arbitrarily define pauses between the pulses that are either "short" or "long" depending upon the time duration between sequential pulses. Perhaps a pause of less than two seconds is defined as a short pause while one lasting longer than four seconds is a long pause.

With this method it would be a relatively easy matter for the operator of the device with tube A on Earth to send simple Morse code messages to the receiving person who operates the device containing tube B on the other side of our galaxy. When a message was completed, the operator of the device with tube B would use its surrounding coil and its attached switch to send Morse code messages back to the person on Earth.

There are, however, problems with this method of instantaneous interstellar communication.

The major problem is that Morse code is not that easy to learn and requires a period of specialized training for a person to become adept at using. It is also not unusual for confusion to take place and words to be missed or misspelled during such communications. And, of course, it is, even though instantaneous, rather slow process of sending messages back and forth. It is for these reasons that none of Earth's current military forces uses this method and prefer to use encrypted digital microwave frequency communications.

Fortunately, it is possible to engage in direct instantaneous audio or speech frequency communication between two remotely separated devices as described above. This, however, requires that the relaxation times for the misaligned or flipped non neutron protons in two spin correlated sister hydrogen nuclei be made much shorter than is required for the simple transmission of pulses that use the time pauses between them to represent dots and dashes as are used in Morse code.

For speech frequencies up to 10,000 Hertz or cycles per second the relaxation time must be reduced to 0.0001 seconds. This is possible to achieve by using very strong end magnets on each tube and exciting the spinning non neutron protons into flipping in opposition to their tube's internal alignment magnetic field by using higher than radio wave frequency microwave photons. With this upgraded device, a person speaking into a microphone on Earth could have a copy of his voice instantly emerge from a speaker on a similar device located on the other side of our galaxy!

Once humanity has finally mastered the construction and operation of anti-mass field generators and begun to construct its own massless UFO like air and spacecraft, it will only be a short time thereafter before explorers will set out to chart the rest of our galaxy. Launching off into the depths of deep space and engaging in journeys that could last weeks and months or even longer if visiting nearby galaxies can be a rather intimidating. For this reason it is important that our explorers be able to maintain regular and instantaneous contact with their bases on Earth and even with each other. Again, the devices described above can be further modified to allow for this.

Imagine that, when the helium nuclei were first split to form *two* separating streams of spin correlated or quantum entangled hydrogen nuclei, a slight change was made.

Instead of having *all* of one of those split streams flow into tube B, let that stream be carefully further subdivided so that it forms a dozen smaller streams that are made to flow into a dozen smaller *separate* tubes each of which is equipped with its own set of end magnets that will maintain the directions of the non neutron proton spin axes and magnetic field directions of the smaller tube's hydrogen nuclei. Thus, instead of a single tube B, we now have a dozen tube B's. However, all of the hydrogen nuclei that were supposed to be placed into tube A will still go into that larger single tube. We now remove all of the tubes from the helium fission chamber and equip each with its own transmission / receiver coil and the associated circuitry needed to allow each to receive and send voice frequency messages.

Now we have a situation where *all* of the hydrogen nuclei non neutron protons in *each* of the dozen smaller tube B's is quantum entangled with 1/12th of the hydrogen nuclei non neutron protons in tube A. If an operator on Earth were to speak into a microphone connected to tube A, then all dozen of the communication devices containing the tube B's would be activated simultaneously and would produce a copy of the sounds being spoken into tube A! This could be useful when it was necessary to issue some sort of common message to a fleet of a dozen of Earth's exploratory craft that might be located in different and very far separated sectors of our galaxy.

It's even possible to send a message to only one of the far flung tube B's. All that is necessary is that an operator of the device on Earth containing tube A first send a prefix code to his message (this code could just be a audio signal of a particular frequency or a certain number of sound pulses in rapid succession) which would then be simultaneously received by all of the tube B's. However, each of the devices containing one of the smaller tube B's would have what is known as a "discriminator" circuit built into it. This circuit only allows the particular device to reproduce an incoming message if that message is preceded by the correct prefix code. Thus, each of the communication devices being carried aboard exploratory craft located in various parts of our galaxy could actually have its own unique telephone number!

Now imagine that, when the stream of split helium nuclei was first divided up inside of our hypothetical helium fission chamber, the single stream that was ushered into tube A had instead been placed into a dozen smaller tube A's. In this situation, each of the dozen tube A's could be installed into its own communication device which, using the correct prefix code, would be able to communicate with any of the dozen devices containing the smaller tube B's throughout our galaxy. In this case the hydrogen nuclei non neutron protons in *each* of the dozen smaller tube B's is quantum entangled with only $1/12^{th}$ of $1/12^{th}$ or $1/144^{th}$ of the hydrogen nuclei non neutron protons contained in any of the smaller tube A's and additional amplification of the audio signal reproduced by any of the dozen tube B containing devices will be necessary. This would then allow any one of multiple human bases on different planets and moons in our solar system to communicate selectively or simultaneously with up to a dozen exploratory craft that might each be located very far from our solar system.

By now the reader may have noticed an obvious problem with the above described instantaneous interstellar communication system. It does not allow any of the distant exploratory craft to communicate with each other and neither does it allow any of the multiple human bases in our solar system to communicate with each other. This is because there is no quantum entanglement between only the non neutron protons in the tube A's or in the tube B's. But, even this problem might be solvable by arranging things inside of the helium fission chamber so that one produces a certain number of tubes each of which contains some non neutron protons which are spin correlated with some on the non neutrons contained in each of the other tubes. The number of tubes created would then be installed into an equal number of separate communication devices each of which would then have its own unique prefix code that would have to be received before any following message could be reproduced by that particular device. Using modern electronic components, such communication devices might be made as small as a toaster and could easily be powered by rechargeable batteries.

Such exotic communication devices would only be intended for use over distances that made regular radio wave or laser beam communication inconvenient because of the long delay between the time a message was sent and a reply received. For communication within a planet's atmosphere radio waves and lasers are certainly acceptable.

After this chapter was completed, I realized that there was another startling possibility for the use of an instantaneous interstellar communication device.

By greatly increasing the strength of the end magnets attached to the tubes containing the spin correlated hydrogen nuclei and then also using even higher frequency radiation to excite them into misaligning with a particular tube's internal alignment magnetic field, it would be possible to greatly reduce the time needed for the misaligned magnetic fields of the non neutron protons to flip back into their normal aligned orientations. That would then make it possible to greatly increase the amount of information that could be communicated between two such devices. In fact, it would make it possible for complex telemetric data to be instantly transmitted between *any* two locations in our cosmos.

Now imagine that, instead of someone on Earth just sitting back and sending a verbal message to another person located diametrically opposite to the sender's position on the other side of our galaxy, the person on Earth had sensors attached to various parts of his body which would then pool their digital data about the forces being generated by those body parts and send that data instantly to an *android* body at a remote location in our galaxy.

That information would then be used to activate various servo motors in the android's body that would cause its corresponding body parts to apply the same magnitudes of forces to objects in its environment. Simultaneously, the android body would have sensors in its body which would detect any reactional forces being applied to its body parts by the objects in its environment that it touches so that the magnitudes of these forces could be coded as digital data and then be instantly sent back to the person on Earth. Once the return or feedback data was received, it would be used to apply forces to the person's body so that he would be aware of the amount of force he was applying to the objects in the environment of the remotely located android body. With this enhanced capability, a person on Earth could actually "feel" the shapes and structural strength of various objects in the environment of the android! Such a virtual tactile sense would also include the ability to detect the temperature of those objects.

This capability, however, need not be limited to the sense of touch. Indeed, it would literally be possible for a person on Earth to "see" stereoscopically through the android's eyes, "hear" stereophonically through its ears, "smell" through its nose, and "taste" through its tongue. Using such a system would allow an Earth located person to, in effect, instantly extend his awareness and ability to manipulate the environment to any portion of our cosmos!

By using such remotely controlled androids to pilot massless exploratory craft it would be possible for humans to explore even the most dangerous of environments. They need not fear any injurious or even lethal magnitudes of android sensory data being fed back into the full body virtual reality equipment that they must wear in order to "become" the androids because the system would be set up to block any such data from affecting the human being should it exceed a certain safe level. Thus, should the android, while exploring some planet near the center of our galaxy, suddenly find himself caught in an avalanche and crushed until he was destroyed, this data would only produce enough force on the human operator's body to make him aware that the android was no longer functional. The human would then simply remove the equipment from his body and be completely physically unharmed.

Using these human directed androids would allow people to engage in such potentially catastrophic endeavors as exploring the interior of the event horizon of a black hole, venturing into the surface atmosphere of a star, or plunging into the life bearing oceans trapped beneath the frozen surfaces of a star system's outer ice moons. Indeed, I've often wondered if some of the humanoid creatures that are described in the UFO reports as moving "like robots" might not, in fact, be these types of remotely controlled machines!

Of course, it is not strictly necessary that such a device be made to look and move like a human being. It could really be any shape one desired. One could just bypass the use of an android body and have the Earth located human "become" the massless UFO itself. That is, the human would come to perceive his body as being the craft.

Perhaps he would control its straight line motion velocity by leaning forward or backward and would then steer the moving craft by leaning to his right or left side. He would be able to "see" out of a set of cameras located on the leading edge of the craft. He could fly through the atmosphere of a planet he is exploring in this manner and then, by squatting, cause the remotely controlled craft to lower down in the atmosphere and land on the planet's surface. The human could reach down to the floor of his room and robot arms would then extend from the landed craft at its remote location. The human would then use his fingers to activate and control fingers at the ends of the craft's extended robot arms and these could be used to pick up such things as rock and soil samples and place them into a storage compartment inside of the craft. Any samples collected would then be quickly returned to Earth for study.

Again, the chief advantage of such virtual exploration of our cosmos is that it can be done without any physical risk to the explorers and, more importantly, it can be done in "real time". If a particular craft is lost during such an exploration, it is merely replaced with a new one and the exploration can be resumed. Every precaution would be taken to assure that whatever mistakes led to the loss of the first craft would not be repeated.

We will not be able to know if the devices described in this chapter are physically possible until humanity has finally developed the anti-mass field generators that make UFO propulsion possible. When that finally happens, as I firmly believe it will, sometime in *this* century, we can expect many amazing new technologies to come into existence almost overnight. There will, of course, be a revolution in transportation and the exploration of our planet and those others of our solar system, but we can expect exciting new discoveries to be made in the fields of communication, atomic and nuclear physics, medicine, etc. Yes, the physics books published at the beginning of the next century will look very much different than the ones students now labor over. These differences will, hopefully, make for a new world in which such evils as poverty, famine, disease, and war are forever banished from planet Earth.

Chapter 12

Why Major Governments Don't
Reveal Their UFO Secrets

Why most of Earth's governments refuse to acknowledge the UFO reality has preoccupied ufologists from the beginning of the Modern Era of their subject which many consider to have begun with Kenneth Arnold's historic sighting back in 1947. There have been many possible explanations offered including that of the skeptics which is that there isn't anything to hide and then reveal because UFOs that are extraterrestrial in origin and actually visiting us from distant star systems simply do *not* exist.

Currently, the most accepted answer to this riddle, which accepts that UFOs *do* physically exist, has divided itself into two major branches each of which has its supporters and detractors.

One branch leans toward unbridled speculation and claims that there is no acknowledgement because the reality of the UFO / alien presence on Earth would just be too shocking for humanity at this time and must, therefore, be suppressed at all costs for fear that humanity would experience a collective and massive nervous breakdown. That would then result in panicky humans committing mass suicide, disobeying laws, and eventually overthrowing the various governments that had kept them ignorant for so long about the truth of the situation. In other words the truth must be suppressed for the "good" of humanity.

Those supporting this branch, lacking undeniable proof of its existence, often embrace the various conspiracy theories involving such things as "men in black", actually government operatives, who "hush up" human UFO witnesses, black helicopters used for the further intimidation of human witnesses, salvaged crashed discs containing either dead, nearly dead, or still living alien entities, secret underground government bases where the UFO wreckage is stored and studied along with the remains of their pilots, working collaborations between government scientists and those of other planets who regularly visit Earth to share their advanced scientific knowledge with us, and even fertile women being impregnated, either naturally or artificially, by aliens so that their embryos can be "harvested" for genetic experiments on other worlds!

The other branch, however, tends to lean toward a more mundane explanation for the lack of acknowledgement of UFO existence and subsequent release of any information collected about them by the world's major governments. It contends that this is because the major governments are well aware of the UFO reality and are covertly studying it in an effort to be able to produce air and spacecraft with the same dazzling aerobatic capabilities as those displayed by our extraterrestrial visitors which, quite obviously, are in advance of those possessed by Earth's currently available air and spacecraft. Those that support this branch simply state that there is no acknowledgement and then disclosure of data because none of the major governments of our planet wants to reveal how advanced their particular knowledge of extraterrestrial craft technology really is since that would then let their enemies or potential enemies also have access to the same information.

If that were to happen then that additional information gained might just be the extra information that a rival government's scientists needed in order to be able to begin constructing the same type of massless vehicles our visitors are using and then equipping them with thermonuclear weapons. Any strategic nuclear weapons system based on massless delivery vehicles that can travel at tens of thousands of miles per hour *through* the Earth's lower atmosphere could easily neutralize enemy targets on the other side of the world from their launch sites and would be very difficult if not impossible to defend against.

The major advantage of such a massless weapon delivery system is that, if properly designed, it can avoid the widespread fires and ensuing environmental damage that would occur following a large scale exchange of conventional missile delivered nuclear weapons. Such an exchange could put so much dust, smoke, and radioactive toxins into our planet's atmosphere that it could eventually destroy all life on Earth either directly or through the triggering of a "nuclear winter" in which all plant life and then animal life died off because insufficient sunlight would be able to reach the surface of our planet. The seldom discussed danger of our present MAD or "Mutually Assured Destruction" nuclear weapons strategy is that, if actually used, it would probably eventually destroy *both* nations involved in the conflict as well as all the rest of Earth's nations that remained "neutral" during the weapons exchange.

Massless UFOs, however, could kill enemy military personnel right in their bases by delivering and then detonating neutron bombs over those bases that lethally damaged the genes of their soldiers' bodily cells and prevented them from performing their normal metabolic functions. Such UFOs could also, using the B rich anti-mass fields discussed in earlier chapters, physically interpenetrate the ground to then detonate low yield nuclear warheads underground where they would collapse and destroy secret subterranean bases and missile silos. Even without using anti-mass fields with B rich character, these massless delivery vehicles could dive at thousands of miles per hour into the surface of an ocean without damage to home in on and destroy missile carrying submarines by detonating low yield nuclear warheads in their vicinity. Any feeble response by a defending enemy government to such a coordinated sneak attack with massless weapon delivery vehicles could be easily thwarted as each of an enemy's surviving launched nuclear missiles was destroyed before it had completed the boost phase of its trajectory that would carry it to its preprogrammed target. In essence, a UFO technology based weapon delivery system allows an attacker to neutralize an enemy nation's nuclear conventional missile based defense system without the need to fear any sort of effective counter strike response or ensuing global environmental damage.

Needless to say, with this kind of awesome capability depending upon which major government first successfully reverse engineers the propulsion technology of our extraterrestrial visitors, it is easy to understand why these governments would want to keep their research into the matter as secret as possible. This secrecy would not just pertain to foreign governments, but also to its own citizens including even the highest ranking political and military members of its own government! Access to the information any major government's elite researchers had gained over the decades concerning UFO technology would be on a strictly "need to know" basis and just wanting the information for the sake of curiosity or to fulfill some campaign promise would certainly not be considered sufficient.

So, in evaluating the two current theory branches that attempt to explain the now decades long major government denial / secrecy concerning the UFO reality, I've reached the conclusion that the truth of the matter is somewhere *between* the two extremes of opinion that these branches represent. What follows now is what I believe that truth most likely to be.

I begin by noting that after several decades of investigating civilian and military UFO sightings, the most that the US Air Force will say on the matter is that they have determined that UFOs do not represent a threat to the citizens of the United States. It is a rather interesting statement in that it tells us absolutely nothing of the nature of the UFO phenomenon or if it even exists at all. It just tells the citizens of the United States (and, by extension, the rest of the Earth) that they need not worry about this phenomenon causing them any sort of harm.

It's obvious to me that statements like this were prompted by the apparently random nature of the sightings and the wide variety of shapes reported for the airborne objects spotted. There does not seem to be any sort of pattern to them that would suggest they represent an imminent attack from space and, in the vast majority of the cases, no harm is done to the witnesses or any property. Indeed, rather than being frightened by the appearance of a UFO, most people find themselves fascinated by it and filled with a sense of awe. Since UFO sightings actually predate by millennia the beginning of the Modern Era of sightings that occurred in 1947, it would seem that, if any of the objects observed had been the vanguard of an attack, then it would certainly have happened a long time ago and there would, currently, be no doubts about the existence of these craft and their operators motives. That, fortunately, has not happened and, most likely, will not happen (but, then again, past performance does not absolutely guarantee future performance).

Of course, I do not doubt for a moment that certain elite groups within the major governments of Earth are well aware of the UFO phenomenon's reality and have been studying the subject intensely since the beginning of the Modern Era and even, perhaps, as early as the World War II years when "foo fighters" were beginning to interfere with the navigational compasses and engine ignition systems of the

Allied bombers as they flew on their deadly missions into the dark heart of the Third Reich.

Initially, it was believed that UFOs were advanced aircraft of earthly origin and, eventually, by the early 1950's it was finally realized that this was definitely not the case. By that time it was realized that some of the objects being sighted were displaying aerobatic maneuvers that far exceeded any that could be produced by any of the then existing or predicted to exist propulsion systems. It was realized that only an extraterrestrial explanation made any sense and, further, that the origin of these craft must be outside of our home solar system. That, of course, meant that the extraterrestrial beings that constructed and flew these marvelous craft had solved the problem of traveling through outer space at velocities far in excess of that of light which is about 186,000 miles per second!

At this point, the major governments could have come forth and announced that it was obvious that *some* of the aerial objects people were reporting were, indeed, advanced extraterrestrial air and spacecraft that were operating in our planet's atmosphere for purposes that were not then understood. This announcement, unfortunately, had a few problems associated with it.

If it was made, then there would be "official" verification that human life was not the only intelligent life in our cosmos. It would have to be admitted by the major governments that there were craft flying around our night skies that were being piloted by creatures that, while humanoid, were not exactly human. They would have to confess that, other than chasing these objects with jets whose pilots occasionally were killed in the pursuit, there was really nothing that could be done about the matter. Indeed, on rare occasion, some of these craft even landed and, using some sort of telepathic mind control, were abducting human beings for medical examinations aboard their craft! These are all, of course, rather disturbing things to accept as true in a world that was still reeling from the nightmare of World War II and was beginning to feel the psychological stresses of living in the "atomic age".

There was also real fear that such revelations could lead to a collapse of the world's various religions.

These almost all teach that there is a supreme and eternal being that created humanity in his image. How could this article of faith be reconciled with intelligent, living creatures that did not look human and, in fact, might be quite repulsive by human standards of beauty? One might then, in light of acknowledging that the UFO phenomenon was real, easily reach the conclusion that all of the world's major religions were just hoaxes and had no validity. That conclusion would then undermine one's belief in an afterlife and some sort of metaphysical salvation into an eternal life beyond the grave. Possibly, such a collapse of religions and the various moral and ethical values they try to foster would lead to increasing conflicts and wars around the globe that could only end with the use of nuclear weapons on a far more extensive scale than was seen in Hiroshima and Nagasaki!

Faced with these realizations, the major governments of Earth made a decision to try to "change the subject" away from UFOs and toward more mundane things such as the economy, the "cold war", and the latest achievements of Earthly science and technology. Changing the subject involved minimizing its importance by ridiculing and occasionally intimidating UFO witnesses and setting up "official" government projects to "look into" the matter.

Initially, these tactics were successful. In the early '50's most of the citizens of the US viewed our government as sort of all knowing and all wise. After all, they had managed in a little less than four years to defeat the military might of both the Nazis and the Japanese in a two front war. If they said that they could explain away 90% of the alleged sightings of UFOs as nothing other than mistaken natural and manmade objects, hoaxes, or hallucinations, then it seemed logical that this "analysis" could be extended to all such sightings if only some additional data were available.

These government sponsored UFO research projects, however, really had, I believe, a far more important purpose.

They provided the elite groups of scientists studying the UFO phenomenon with easy access to any witnesses that, along with a sighting report, also were able to produce some sort of tangible evidence such as photographs, films, or materials that might been left behind by a hovering or landed craft. That material would then be quickly passed on to those studying the subject and would only very rarely ever be returned to a witness. He or she would simply be told not to discuss the matter with anyone in the event that the object observed was some sort of secret government air or spacecraft. They might also be cautioned to remain silent because the object sighted might be an enemy aircraft and we did not want our enemies to know that we had any information about their activities in our skies.

Perhaps 99 out of 100 witnesses would simply accept this need for silence and maintain it. When an individual came along who wanted to share the details of his sighting with his fellow citizens, that person would be intimidated into remaining silent by strange "men in black" type characters that would show up at his home often at night when he was along. They would then "hint" that any further disclosure could have dire consequences for the witness and his family. Usually a single visit from one of these characters was enough to permanently silence a person.

Of course, all major governments must have an exit strategy for their UFO projects which is intended to end their further involvement in it. This usually takes the form of a final "scientific" study in which it is concluded that there is nothing of value to be learned from any further study of UFOs and that the objects being reported do not represent any sort of threat to the government's citizens. They then publish what they claim are all of the cases they've collected over the years, but later investigation will inevitably indicate that there are always a few cases that seem to have mysteriously disappeared from the published collection.

Of course, such published studies really never settle the matter because the UFOs just continue to be spotted, photographed or filmed, and reported to various civilian UFO research groups. They then collect and preserve the data and make the most reliable and strange cases available to their own researchers for analysis. These groups often do a far better job in collecting and evaluating the data than do the government UFO investigation projects. But, the civilian UFO groups are always hampered by not having the same quality of evidence that major governments possess that comes from their own trained military observers who often have the latest methods for recording the incidents available to them.

At the present time, a sort of stalemate has been reached with regards to how the major governments are handling the huge store of high quality UFO data that certain elite groups within them possess.

Obviously, these elite groups are not releasing this information because they feel that any official verification of its existence would just be too unnerving for the average citizen to handle and might even result in the overthrow of the major government's political and military system! Also, as I mentioned above, there is now a "UFO technology race" underway between the elite scientific groups of the world's major governments to see which can reverse engineer the propulsion technology displayed by our extraterrestrial visitors' vehicles and incorporate it into a system that would then allow one nation to launch a "first strike" on an enemy with next to zero probability of any effective response being made or of any significant damage being done to the Earth's atmosphere and biosphere.

This stalemate will, most likely, continue until one of two possible events occur.

The revelation of a major government's involvement in the collection and analysis of UFO data will probably be immediately made if one of the extraterrestrial races operating some of the UFOs being observed in our planet's atmosphere decides to openly land one of their craft and make contact with representatives of that major government or a strategic massless weapon delivery system, reverse engineered from recovered crashed UFOs, is actually used in order to thwart a "standard" nuclear missile attack on the major government that covertly developed the new weapon delivery system.

With official and ongoing contact being made with an alien space race, the release of a major government's knowledge of their technology would help to lessen the feelings of inferiority that the government's citizens might feel about the event. Thus, the government could claim that they knew all about the alien activity for decades and had also been studying it that long in order to develop an adequate defense against the intruders of our world just in case they proved to be hostile. Knowing one is on a more or less even level technologically with "visitors from the stars" can help to lessen one's anxiety about their sudden open appearance on our planet and, of course, on the television screens by which most of us view that planet.

In case a major government uses their UFO technology reverse engineered massless weapon delivery system to successfully stop an aggressor's attempt at a nuclear first strike, then the elite scientists who made it possible would be under enormous pressure to explain how they performed such an apparent miracle. Once the system was openly used, as happened with the first nuclear weapons used to destroy two Japanese cities in 1945, the whole world would quickly become aware that such a system exists and, most likely, those powers which still did not possess it would dramatically increase their efforts to obtain the same system. Unfortunately, just as such a system can be used for defense, it can also be used for offense and that detail will then be used to rationalize why every nation on Earth will need one in order to prevent the major government with the system from imposing its will on all of the other nations of our world.

As soon as the big secret is finally revealed that one of the world's major governments actually possesses the technology displayed by our interstellar and even intergalactic visitors, that nation will tend to dominate the rest on our planet. This domination, if done by the United States, will probably be benevolent in nature and serve to influence the way the rest of the nations conduct such things as their politics, laws, and economies. In a best case scenario, this domination could finally lead to a truly unified world in which there was only a single global super nation with what were previously sovereign nations being reduced to states within that nation. Eventually, everyone will benefit from having the same political system, laws, and currency. With it the various inequalities that currently exist between the peoples of Earth will be eliminated and everyone will have the same chance of reaching their full potential for creativity and productivity.

With the formation of a global super nation on Earth, the next priority will be the defense of our home planet.

Using Earth's new massless vehicle technology will allow the people of Earth to stop the previously unregulated visitations to our planet by extraterrestrial craft unless such visitation is authorized by the people of Earth. Once our home planet is secured, this regulation will then extend out to our nearest neighbor in space, our moon, and then, finally, out to the rest of the planets in our solar system. And, of course, as we humans venture out into the surrounding galaxy, we will find that our visitations to nearby inhabited star systems will also be subject to regulation by *other* beings. Perhaps, if we demonstrate that we intend to be a peaceful race of beings, we will finally be allowed to join some sort of local star system group or federation whose members agree to adhere to certain common laws as they conduct visits and trade with each other. No doubt, this membership will be one of the most exciting times in the history of the human race!

So what can I say with absolute certainty about what the major governments of Earth currently know about the UFO phenomenon?

In my own research into this fascinating subject (which spawned a trilogy devoted to it), I managed to make significant progress after about only twenty years and by using openly available data. One should then consider what must have been achieved by the elite scientists of Earth's major governments that have been studying the phenomenon for over 60 years and have had access to hundreds or even thousands of good quality, close up films and videos of these craft.

I have, therefore, little doubt that, by carefully analyzing that material and the recovered debris from dozens of crashed craft, some of the major governments of Earth have *already* successfully reverse engineered the propulsion technology of the UFOs and are now busily at work constructing the massless weapon delivery system I described above. The reader can also be quite sure that the level of security on all of this must be nothing short of mind boggling.

Those who have access to the extraterrestrial materials being studied probably have to agree that they will either spend a significant portion of their lives living in "secure" underground research facilities (which means that they are like maximum security subterranean prisons) or, if they must physically leave the research facility to resume a normal life again on the surface, then they must have previously agreed to have their memories erased using various drugs so that they will not remember ever having worked on the project! Such elaborate precautions might seem bizarre to the reader, but they would be deemed acceptable and necessary because the secrets being guarded could put the entire future of a nation at risk were they to be prematurely divulged by a single disgruntled scientist who managed to escape the UFO reverse engineering research project with actual undeniable evidence of its existence. And, in the unlikely event that someone actually did escape such a program, he would probably be quickly captured and subjected to memory erasure. If that proved difficult or impossible to do for some reason, he might even be "terminated".

There is, however, still some hope that the secrets of UFO technology can finally be revealed and made available to the average citizen of the world. That hope involves the research being conducted by the various civilian UFO research groups and the scientists and engineers that are members of them. I have tried to present my own research into the subject as best I could with the intention of spurring interest in it by promoting the idea that UFOs are not some mysterious and totally unknowable phenomenon. I believe I've done much to show that these craft are as real as automobiles and operate by physical principles that are only a little in advance of the ones contained in our late 20th and early 21st century physics textbooks. Indeed, I fully believe that it will soon be possible for the anti-mass field generators that make UFO propulsion possible to be duplicated and then incorporated into earthly air and spacecraft in order to benefit the average human inhabitant of our planet.

Should that happen, then the elite scientific teams within the major governments that currently possess the best UFO data might decide that they can finally release the full nature of what they've been up to during the last 60 or so years during which they were quietly collecting and analyzing that data and to what ends they intended to it. We would then finally be able to see the preserved remains of the various species of extraterrestrial pilots that had been recovered and the wreckage of their actual crashed craft. We would also be given a general idea of how extensive the new massless weapon delivery vehicle system was and what were its capabilities. By the time this occurs, most of a covert UFO research program's original elite scientists would be dead and could not be held accountable for the secrecy in the event that it was deemed unlawful in some way. For the average person such revelations would, because of the length of time they were spanned, seem like just so much interesting history which would not make much of a difference in his daily life.

Of course, there is also the possibility that these elite government research groups, once the defensive system was finally constructed and made operational, would then just destroy all of the data and materials that they had collected. In effect, they would actually erase history and, perhaps, only a few rumors would remain about it told by an occasional dissident scientist who had been involved in the latter stages of the research program. His testimony would be accepted by some and rejected by some. Without any tangible proof of his claims, his story would just be another colorful tale told in a future book on the history of the UFO phenomenon on Earth.

Would the elite scientists working in the ultra top secret UFO research programs of the world major governments actually do something like that? I believe that there is a good chance that they would.

Erasing their history of activity would then allow the average citizen of our planet to believe that humans were the ones who had really determined the ultimate nature of the UFO phenomenon based on their *own* observation based research and then used the resulting advanced technology achieved to enhance all aspects of life on Earth. The world might eventually learn about the major governments' new massless weapon delivery systems, but would then assume that they were all developed *after* civilian researchers had unraveled the secrets of UFO propulsion systems. If done carefully enough, a hundred years from now the average person might be completely unaware that the world's major governments had had any interest in the subject of UFOs and that their technology at all during the second half of the 20th century or any part of the 21st century.

And perhaps, this is the best reason that the world's major governments have decided to keep their research into UFO technology as securely secret as possible. If it were revealed that these governments now possess massless vehicle technology that had been reverse engineered from crashed UFOs, then for the rest of Earth's history humans would realize that they really did not do it all on their own, but, rather, sort of cheated by having the answers to the mystery handed to them as a result of some extraterrestrial race's unfortunate loss of one of their craft and its crew. Perhaps, in true paternalistic fashion, the elite UFO researchers within the Earth's major governments have already decided to avoid any disclosure of what they currently know so that their civilian populations will have time to figure out the technology of the UFOs for themselves based on whatever evidence is publicly available.

While I can see the positive aspects of this justification for total secrecy, I can also see some serious negative ones to it.

For example, it just might be the case that medical technology based on the ability to negate the normal mass and inertia of matter and change its electrostatic properties could be saving countless lives right now. Being able to control the electrical properties of matter in this manner might also make it possible to economically feed and house the billion or more of Earth's human inhabitants who currently live in squalor and desperation. Such advanced physics might, indeed, make possible an entirely new world where prosperity and freedom were finally available for all. One might then wonder why the world's major governments' needs to develop yet more weapon systems should take priority over all of this good that could come from full disclosure of the data along with an enforceable international agreement to never use it for military purposes.

As I conclude this chapter, I will leave it to the reader to decide how the history books written at the end of this century will describe the UFO phenomenon and the role that the world's major governments played in it.

Chapter 13

Playing with Infinity

About a decade or so ago I visited a large local department store with a friend and somehow found myself wandering around in their toy section. Out of curiosity I decided to do a short inspection of what the latest toys for kids were like and see how they compared to what was available when I was a toy buying kid during the late '50's and early '60's.

I discovered that, with the exception of the introduction of high tech video games, things had not really changed that much over the decades. There were the usual dolls for girls and toy vehicles for boys. Now, however, these were becoming more and more complicated by the addition of microchips intended to make them sound and move in a more realistic fashion. And, of course, practically all of the board games I had played as a youth were still available except that now they cost about five times as much!

While inspecting the game aisle, I happened upon a section that contained various puzzles and "brain teasers". My eyes soon fell upon an interesting little device that had a rather spectacular heyday back in the early '80's. I am referring to the little puzzle cube known as "Rubik's Cube". I had seen these over the years at many of the local flea markets I had visited and, on occasion, had picked one up and played with it for a few moments. Attracted by the bright colors of the newly manufactured cubes at the store, I decided to buy one and see what I could do with it.

If it is possible that any readers have never heard of Rubik's cube, then let me very briefly describe its history and features.

Basically, it is a small plastic cube that measures slightly over two inches on each of its edges. Each of its six larger square sides is a different color and each of these sides is further subdivided into nine smaller square faces of the same color that are arranged so as to form a 3 x 3 array.

There are six smaller square faces that are at the centers of a cube's six larger sides. These six smaller center square faces are each of a different color and always remain fixed in position with respect to each other. Surrounding the single smaller center square face of each of the cube's six larger sides so as to form a cross pattern with its center square face, there are four edge pieces each of which is made up of two smaller square faces at right angles to each other with each of these faces being part of one of a cube's two adjacent larger square sides. Finally, there are the eight corner pieces on the cube each of which is formed from three of the smaller square faces all of which are at right angles to each other with each of these faces being part of one of a cube's three adjacent larger square sides. As a result, each cube contains a total of fifty-four smaller square faces on its six larger square sides that are attached to twenty-six underlying plastic pieces. Six plastic pieces each carry one of a cube's smaller square center faces, eight plastic pieces each carry three of a cube's corner forming smaller square faces, and twelve of the plastic pieces each carry two of a cube's edge forming smaller square faces for a total of twenty-six plastic pieces carrying fifty-four smaller square faces).

This gadget was invented in 1974 by a Hungarian Professor of Architecture and Design named Erno Rubik. By 1980 he managed to interest a toymaker in the design and within a year of its introduction it became one of the world's best selling puzzles. What makes it unique is that it contains an internal cylindrical mechanism which holds its twenty-six face carrying plastic pieces together while still allowing each of a cube's six larger sides to be rotated about its smaller center face independently of the cube's other five larger sides! The cube feels very comfortable in the hands and its various face carrying plastic pieces can be moved very smoothly so that each of a cube's edge or corner pieces can be quickly placed into any of their other allowed positions with various orientations. Since the six center pieces can only *rotate* along as one of a cube's larger square sides is rotated, this means one can really only *translationally* move the remaining twenty face carrying plastic pieces around the cube so as to change the orientations of the colored faces attached to them.

The basic idea of this puzzle is disarmingly simple. When you first open the package it comes in, you will find that each of the cube's six larger square sides is made up of nine smaller square faces of the *same* color (on the one I bought the six larger square sides each have nine smaller square faces that are colored red, yellow, orange, green, blue, and white). You can then grasp the cube in both hands, close your eyes, and proceed to randomly twist and turn its various "layers" of face carrying plastic pieces with respect to each other (one can think of a cube at any moment as consisting of nine distinct layers that can be rotated with respect to each other). After about a minute or so of doing this, you will then open your eyes and discover that the smaller colored square faces on the cube are completely scrambled. Now, all you have to do to "solve" this little puzzle is twist and turn the cube's various layers again (with eyes open this time!) until you finally move its twenty moveable face carrying plastic pieces about in just the right way so as to properly orient them and thus unscramble the nine smaller colored square faces on each of the cube's six larger square sides and thereby restore the *entire* cube to its original appearance of having only nine smaller square faces of the *same* color on *each* the cube's six larger square sides.

Certainly sounds simple enough.

In reality, however, it is one of the most difficult puzzles ever invented! Probably about 99+% of the people who purchased this item during the early '80's never did manage to solve it. My efforts almost came to a similar fate. After about an hour or so of fiddling with my cube, I did manage to complete one face which then became the top layer. Over the course of the next few *days* I attempted to move on to the next lower middle layer and soon discovered that efforts to move its edge pieces into their correct positions and orientations on this layer tended to cause correctly placed pieces in the completed top layer to move *out* of their positions! After about a *week* of this frustration, I was on the verge of auctioning off my cube on eBay when I finally managed to solve it.

Solving my cube required that I obtain a simple step by step solution. Fortunately, there are websites that offer solutions for the 3 x 3 x 3 cube which are based on several books that were written about the puzzle during the '80's. I won't go into the details of the solution here except to say that it requires one to work his way down from the top layer of the cube to the bottom layer using a variety of complex manipulations to move each plastic piece one at a time and, if necessary, rotate it once it is in position so that its faces' different colors each lie on the correct side of the cube. If one makes a single mistake while performing these manipulations, then there is the very high probability that the entire cube will wind up so scrambled again that it will be necessary to start over from the beginning with whatever color one chose for the top side of the cube!

So, while I admit that I like the primary colors and feel of my cube, I must confess that I did not really enjoy it as a puzzle. I found it to be too complex an object for which to readily visualize a solution. In order to solve it, one is reduced to memorizing a variety of standardized manipulations of the various cube layers so that its various fifty-four smaller colored square faces will all eventually be assembled into their correct locations on the sides of the cube. (Despite these drawbacks to this "brain teaser", however, learning to solve it and doing so regularly might be a good way for people to stimulate and strengthen the memory centers of their brains and thereby help ward off the natural decline in cognitive abilities and memory that comes with advancing age.)

Of course, the difficulty of Rubik's cube has to do with the enormous number of possible combinations into which its various smaller colored square faces can be configured. To be precise, a 3 x 3 x 3 cube with twenty-six visible smaller square faces of six different colors being carried on the outsides of twenty unseen movable plastic pieces can be configured into exactly 43,252,003,274,489,856,000 possible combinations which we can just round off to 43.252+ million trillion combinations!

Anyway, my experience with this interesting little invention did get me to thinking about large numbers in general. In particular, it got me to thinking about the concept of infinity which is what the remainder of this short chapter will deal with. First, however, let's very briefly review some very basic mathematical concepts.

Numbers are the way that we describe the magnitudes or sizes of groups of objects and these magnitudes are represented by symbols called "numerals". In the world we live in, we quite routinely assign numbers to two or more groups of objects with some quality in common and then use these numbers to decide which of the groups is biggest in membership. From the Arabic number system, we received the numeral

"0" which is the number zero. It tells us that a group of this magnitude contains *no* members. Sometime during the Middle Ages, "negative" numbers were invented. They tell us that the number of members in a group is *less* than zero. Many beginning students of mathematics are disturbed by the concept of groups of object with negative magnitude when they are first introduced to it, but it merely indicates the magnitude of deficiency of members of a group. The more deficient in members a group is, the more members that will have to be put into the group to *increase* its membership back to zero!

For example, if you had a bowl containing three apples, then you would represent the number of apples in it with the numeral "3". If you ate all three apples, then the number of apples in the bowl would be represented by the numeral "0". However, suppose over the course of the next three days, you borrowed three apples from a friend and ate them all. Now you could consider your bowl to contain negative three apples that you would represent by the numeral "-3". If you found three apples on a tree and decided to add them to your bowl, you might well find, perhaps a few hours later, that you have zero apples in your bowl. Why? Because the friend whom you borrowed the three apples from days ago might come along and take back the three apples you owed him. Thus, we see that the 3 apples you found and placed in the bowl actually added to the negative three apples already there to yield a total of zero apples in the bowl.

The above example is, of course, a simplistic one. Negative numbers, however, are critically important in the science of physics. When two forces acting on an object are in opposition to each other, the magnitude of one of them can be represented by a negative number and this can then be used to find the *net* force acting on the object and it is this net force which will determine what kind of motion the object will undergo.

It is also possible to use a straight horizontal line to represent the various numbers we use. On this line an arbitrary point is selected to act as the reference point and is said to represent the number zero (this point is also sometimes referred to as the "origin"). A fixed length is then used to mark off the regular numbers we normally use on this line with the marked off numbers being placed to the *right* side of the zero point's position on the line. These regular numbers are also referred to as "positive" numbers. On the horizontal line to the *left* side of the zero point's position, we can use the same fixed length to mark off the so-called negative numbers we use in mathematics. Thus, the numbers we encounter as we move along the horizontal line to the right of the zero point are 1, 2, 3, 4, 5, etc. and the numbers that we encounter as we more along the horizontal line to the left of the zero point are -1, -2, -3, -4, -5, etc.

The above described horizontal line which is used to represent the numbers we use is referred to as the "real number line" in mathematics. It is also possible to construct another line of numbers at right angles to the real number line and passing through its zero point. This second line is referred to as the "imaginary number line" and represents the imaginary numbers used in certain advanced forms of mathematics. Its origin will be at the same point as the origin for the real numbers. We need not delve into this extra type of number here, however.

The interesting thing about this way of representing the numbers we use is that the line used can then be *indefinitely* extended in *both* directions. Thus, there is no limit to the magnitude of a group of objects or the number that represents that magnitude, whether positive or negative, that can be represented with these lines!

However, in the macroscopic world that humans normally inhabit and are locally aware of, we have no need to represent the magnitude of a group of objects by a line of infinite length. This is because practically everything humans observe is finite in size, duration, or strength. For examples, consider that we know that we own so many cars, houses, or dollars in the bank. We will live so many seconds between the times written on our birth certificates and death certificates. That there are so many seconds, minutes, and hours in a solar day. And, that the Earth and the other planets in our solar system have a fairly constant size, mass, orbital period, etc.

The point of all this is that we are, from the first time we become aware of the limits of our playpens, conditioned to think in terms of *finite* numbers. After a while, it becomes natural for us to believe that everything in the universe is, in fact, finite in nature. Because of this, ancient peoples developed religions that described creations of humanity and our world which occurred at some specific time in the remote past and which also predicted catastrophic ends to humanity and the world which would occur at some specific time in the future. Astrologers talked about how the "fixed" stars in the stationary vault of the heavens could determine the unalterable destiny of humanity. Even the early mathematics of geometry dealt with various shapes of fixed and finite volume.

In thinking these matters over, an interesting thought occurred to me. What if, in reality, the true nature of everything is not finitude, but, rather, *infinitude*! Let us now continue to pursue this idea to its ultimate and quite surprising conclusion.

In the science of mathematics there is a symbol used to represent infinity. It looks much like the Arabic numeral for the number 8, only it is lying on its side and is represented by ∞. In mathematics one learns that if one divides a finite number by another variable finite number which is then allowed to dwindle in magnitude toward zero without quite reaching it, then the quotient of the two numbers will increase toward infinite magnitude, but, again, without quite reaching it. The implication of this is that any finite number divided by zero must equal infinity and this can give one the impression that infinity is just another number.

However, the reality of the situation is that infinity is not a number at all, but rather a *concept* of limitlessness. It is a concept that is totally foreign to the way humans perceive the world around themselves. Yet, it is a concept which ultimately is the final reality of all that exists, ever existed, or ever will exist!

In a past work (see the chapter titled "Our Extraterrestrial Universe" in my book *The How and Why of UFOs*) I suggested that we humans actually inhabit only one local universe that is an infinitesimal part of an infinite array of such universes that make up our present "multiverse". At approximately the same time, each local universe starts off as a "Cosmic Egg" which is a solar system sized Black Hole that contains a sphere made of solid neutronium as its central "singularity". This singularity has a finite volume and mass.

These Cosmic Eggs slowly grow over the course of tens of billions of years by drawing in dust and gases from incoming dying galaxies and, due to their immensely powerful gravity fields, crushing these materials down to create and add more neutronium to their surfaces so that they grow via a process of slow accretion. When they reach a certain critical size, however, the Cosmic Eggs become unstable and a mechanism is initiated whereby their gravity fields are suddenly weakened. At this point, due to the enormous pressures inside of their cores, the Cosmic Eggs all "hatch" at about the same time with a massive explosion taking place in each one that causes it to release an expanding cloud of neutrons in all directions. After several billions of additional years, these clouds eventually form all of the galaxies in each of the infinite number of local universes that make up that particular multiverse.

Over the course of a hundred billion years or more, the galaxies that form any single local universe move steadily outward and away from the location of the Cosmic Egg detonation that created them. Eventually, all of the various energies stored in the outermost boundary layer of expanding galaxies in any local universe are dissipated as they radiate off into space in the form of electromagnetic energy and these galaxies begin to die. Finally, these galaxies begin to collide with the *incoming* dead and dying galaxies from *other* neighboring local universes and, eventually, all of their matter coalesces to form a new Cosmic Egg which then continues to grow as it draws in more and more galactic debris with its immensely powerful gravity field.

A point in time is finally reached when all of the matter and energy of the multiverse is again locked up within an infinite collection of Cosmic Eggs. As a result, for a period of, perhaps, tens of billions of years, the multiverse is completely dark. Then the infinite set of Cosmic Eggs begins to simultaneously hatch and a brand new multiverse is created. However, all of the physical laws in this new multiverse will be identical to those in all of the other previous multiverses that have ever existed of which there is an infinite number.

The above is simply a "Big Bang / Big Crunch" or "Oscillating" model for the cosmos, but it postulates that the processes involved are infinite in *both* space and time! That is, the process is actually occurring throughout infinite space, has been in operation from the infinitely remote past to the present, and will be in operation from the present into the infinitely remote future. This is an entirely natural process and, in fact, is the *only* possible process that can exist!

We see from this model that the concept of infinity is not just a metaphysical or theoretical one, but rather the natural state of affairs everywhere throughout eternity. Since the multiverse is infinite in both space and time, one wonders what other strange concepts might also be possible for our infinite multiverse. Let us now explore some of these.

Upon contemplating these matters, one immediately comes to the conclusion that, with an infinity of local universes making up our present multiverse, there must be a very large number of universes which are identical to our own local universe. By identical, I mean that these other universes will have formed from a Cosmic Egg that was *exactly* the same as the one that hatched to form our present local universe. Their Cosmic Eggs would have contained exactly the same finite number of neutrons as ours and the motions of all of those neutrons with respect to each other would have been exactly the same as those that existed in our Cosmic Egg about 14 billion years ago. This would guarantee that when those identical universes' Big Bangs occurred, the resulting galaxies that formed in each of them would be the same as the ones in our local universe and would contain exactly the same number of stars, planets, and living creatures. Additionally, the galaxies in each of those other universes would all have been moving in precisely the same way with respect to each other.

In fact, not only is the number of these other universes which are truly identical to ours large, but it must be infinite! If this is the case, then as I sit here writing this chapter, there must be an infinite number of other authors with my name "out there" who are, at exactly this same instant, working on the exact same chapter that you are now reading! They all live on a planet that is identical to "our" Earth and which has the same countries, oceans, peoples, and history as does ours. Indeed, every thought I am having or deed I am performing is, at the exact same instant that I have or do it, being repeated throughout the infinite distances that surround us! Not only is this true for the author, but it applies to the thoughts and actions of every living creature throughout our local universe.

One then might wonder if it would ever be possible for a person in our universe to travel over to one of these other universes which is identical to ours so that he might be able to meet himself. Unfortunately, this possibility is extremely improbable for a variety of reasons.

First, because of the huge number of neutrons in the solar system sized super black holes that compose a Cosmic Egg, the probability of two identical universes forming near each other (like within a few hundred *billion* light-years of each other!) is very low. Therefore, our nearest twin universe might be trillions or even hundreds of trillions of light-years away. Even after we of Earth finally manage to duplicate the mass negating technology that allows UFOs to defy the laws of gravity and inertia and greatly exceed the velocity of light, we may never get beyond the ability to travel to only the galaxies in our own local universe.

Secondly, even if, by an incredible stroke of luck, an identical universe to ours was very close so that we could just barely reach it with a massless spacecraft, we would be trying to meet people who had the exact same idea we did. As we landed on their Earth to meet them, they would, at the exact same time, be landing on the Earth of yet another universe identical to theirs (that universe, most likely, would not be our one). We would find that they were not at home just as they would discover that the people they were visiting were likewise not home.

The implication of the above situation is that, although apparently random in the distribution of its energy and matter, there must be a subtle underlying symmetry to the multiverse. While there are an infinite number of *different* universes in our multiverse, the individual universes that form a *particular* subset of *identical* universes must be symmetrically arranged with respect to each other throughout the multiverse. The most logical arrangement for the identical universes in any particular subset would seem to me to be a simple cubical array distribution for the member universes of the subset. Thus, if one was to represent a very large section of our multiverse by a large cube which was made up of many smaller cubes that were packed closely together, one could then imagine any subset of identical universes to be represented by the points where the corners or vertices of the smaller cubes touch within the body of the larger composite cube. If this cubic array of vertices was then, as a whole, shifted a bit in one direction or another, then the new cubic array of vertices might represent another and *different* subset of identical universes. Again, it must be remembered that each subset of identical universes will contain an infinite number of universes.

Not only are any infinite subset of identical universes symmetrically arranged with respect to each other, but they must also be identically oriented with respect to each other. We can further extend this analysis by postulating that each individual universe of an infinite subset of identical universes is also surrounded by the *same* local universes, each of which is a member of another infinite subset of identical universes. Thus, we see from this that while there is an infinitude of different kinds of infinite identical universe subsets in our multiverse, they are all sort of "locked" into a particular, though enormously complex, relationship with respect to each other.

However, each multiverse which arises between the hatching of its initial infinite cubical array of Cosmic Eggs and its final infinite cubical array of Big Crunches (which then goes on to eventually form the next infinite array of Cosmic Eggs) is unique. If the Cosmic Eggs are arranged with respect to each other as was suggested above when they all hatch, then the expansion of any local universe will eventually, over the course of a hundred of billion years or more, cause its expanding outer front of dying galaxies to finally encounter and begin merging with the similarly outwardly expanding material of twenty-six other local universes that surround it so as to finally create eight new Cosmic Eggs that are arranged at the corners of a cube with respect to each other. Thus, each new Cosmic Egg that is formed during the Big Crunch phase of a multiverse will be different from the previous eight hatched Cosmic Eggs that contributed material to it. Interestingly enough, the total number of nearest surrounding universes whose expanding outer fronts of debris interact with any particular expanding local universe is exactly equal to the number of face carrying plastic pieces found on a Rubik's Cube (i.e., twenty-six)!

We see from this cosmic mixing process that each multiverse will be different from the one that preceded it and will, in turn, give rise to yet another different multiverse. I suspect that, because each multiverse is spatially infinite and, therefore, contains an infinite amount of matter, this will guarantee that throughout eternity no two multiverses will ever be identical.

It is also important to remember that, while the cosmos consists of an infinite parade of multiverses, the major differences between any two of them are really just cosmetic. While there will be great variation in the numbers, positions, and shapes of the galaxies that help form a particular multiverse, the fundamental laws of physics and chemistry will not vary from one multiverse to the next. Thus, the physics books that were valid in any multiverse will still be valid in the next multiverse and will remain so for eternity.

There are many interesting and perplexing questions that arise when one attempts to conceptualize the ultimate nature of our infinite cosmos.

For example, one may readily acknowledge that our present universe is merely one member of an infinitely large subset of local universes which are perfectly identical to ours and that, therefore, at this *exact* instant there must be an infinite number of identical humans out there for each person now living on our Earth. If one further assumes that all of the infinite number of Cosmic Eggs from which each of these identical universes hatched did so at the exact same instant which would seem to be necessary in order to make them all identical, then he may further acknowledge that every current thought and action of his or her infinite number of twin humans out there in this infinitely large subset of identical humans is also perfectly synchronized. But, is it possible that there are also other identical and infinitely large subsets of local universes wherein all of their twins of our Earth's humans are now, at this instant, a few minutes *behind* us or a few minutes *ahead* of us in their thoughts and actions?

In thinking it over, it seems plausible that some temporal desynchronization between otherwise identical infinite subsets of local universes would be possible. This is because, while a multiverse comes into existence rather quickly, it would seem improbable for a multiverse's infinite number of Cosmic Eggs to *all* hatch at the *exact* same instant. Perhaps there could even be as much as a desynchronization of a few billion years.

However, I think if desynchronization began to stretch into the tens of billions of years, then it would not be possible for a multiverse to have a distinct lifespan to it. One would have a situation in which *all* phases of a particular multiverse's lifecycle were *simultaneously* present. At any instant there would be an infinite number of Cosmic Eggs detonating with Big Bang explosions, an infinite number of universes would be undergoing expansion and the inevitable formation of galaxies, star systems, planets, and lifeforms, and an infinite number of universes whose outer expanding fronts of dying galaxies were colliding with those of surrounding local universes and collecting to form the super black holes that become Cosmic Eggs. If this was the case, then there would only be a *single* eternal multiverse that has existed forever, exists now, and will exist into the infinite future.

It would seem that any extensive desynchronization present in our multiverse would, if it existed, eventually tend to automatically extinguish itself as the multiverse became cyclic. It is almost as though the infinitely large cubic array of Cosmic Eggs that repeatedly give rise to each new multiverse was like a marching army. In order to advance as smoothly as possible, all of the soldiers in a marching army must move in the same direction and with the same speed. If half of the soldiers suddenly slowed down, sped up, or changed direction, then the army's advance would be disrupted as soldiers began colliding with and tripping over each other. To avoid this, each solider becomes aware of the motions of the soldiers surrounding him and adjusts his own motion accordingly so as to conform to theirs.

Perhaps much the same occurs with a multiverse. A Cosmic Egg can not fully form and hatch with a Big Bang explosion until it has gathered enough galactic debris into itself from its expanding neighboring local universes. But, these neighboring universes will not be able to contribute that material until their Cosmic Eggs have hatched and their Big Bang universes have sufficiently aged so that they will have an outer expanding front of dying and dead galaxies to provide the material. Thus, we see that there is a sort of self regulatory mechanism involved in this process that, I believe, will guarantee that our multiverse is, indeed, cyclic in nature with an overall lifespan that has a fixed number of years to it.

So, obviously, I tend to lean toward the view that our multiverse is cyclic and, hopefully, one day in the next few centuries we will be able to definitively determine which is the case. With the development of massless spacecraft capable of achieving enormous hyper-light velocities, we will be able to send probes out that can begin to explore the various other local universes that surround the current expanding Big Bang universe in which we are located.

If those surrounding universes are, more or less, in the same state of expansion as ours, then the probes could travel outward to even more distant local universes and check on their phases of evolution. Again, if these observations and those of even more distant local universes prove to be, more or less, in the same state of expansion as our home universe, then that would seem to be very strong evidence that our multiverse is, in fact, cyclic in nature and that we are dwelling in the midst of a never ending *series* of spatially infinite multiverses each of which is finite in duration.. If, however, the probes report that the local universes surrounding our own are in various states of evolution with some having Cosmic Eggs in the process of growing, some having Cosmic Eggs that are undergoing Big Bang explosions, and others being post Big Bang expanding and evolving universes, then that would be strong evidence that we only inhabit a single, eternal multiverse that is static in nature

Some readers may wonder if our current infinite multiverse allows for the existence of an infinitely large subset of local universes wherein their present lives, as they know them here on "our" Earth, somehow followed a different course. For example, could there be an infinite number of any of our currently living earthly humans out there who, in their immensely distant existences, managed to escape some bad or tragic situation or who, on the other hand, managed to achieve some great thing by overcoming great odds against it.

The answer to this riddle is probably "yes".

Just as it is possible for a small amount of temporal desynchronization to exist in even a cyclic multiverse, it should also be possible for a small amount of *structural* desynchronization to also exist without it seriously changing the identities of the infinitely large number of human twins out there. For some of these other humans, the slight structural discrepancies in their environments will serve to favor the attainment of their goals while for others different structural discrepancies will serve to hinder the attainment of their goals. Perhaps, statistically, half of all of those other humans out there will lead lives better than their counterpart on our Earth and the other half will lead lives worse than their counterpart on our Earth. Some will have incredibly successful lives while others will have incredibly tragic ones. Most will have lives that are near to some statistical "average".

Even for those humans with less than average lives, there is always hope that things will improve for them even if this is not to be their final destiny. This hope *always* exists because it is always impossible for us to know, with certainty, exactly what our future life will be like. And it is hope that keeps a person functioning despite the adverse circumstances in his life. Thus, this ignorance about the exact nature of our futures and the hope that ignorance engenders is a kind of blessing that I believe most people would really not want to give up. Thinking one will live to the ripe age of one hundred years is much preferable to knowing for *certain* that one will be dead at fifty years of age

This brings up a final metaphysical matter that I want to briefly treat before bringing this chapter to a close. Mainly, the question is whether or not it would be possible to change the future *if* one had some sort of magical crystal ball that could show one exactly what that future would be.

The answer to this question is that such avoidance would *not* be possible! Surprise! The reason is that if one truly had a crystal ball that showed future events in perfect detail, then the picture it delivered would *already* have taken into account one's viewing of the situation and any subsequent efforts to change it! Thus, if the crystal ball showed a person being hit and killed by a bolt of lightening at a certain time and place, then somehow external events in the person's life, over which he had no ultimate control, would come into action in such a way so as to guarantee that he was at the designated place on time to fulfill his destiny. Quite fortunately, such crystal balls do not exist and probably never will. They would not change one's life in any way, but would only burden one with additional fears and worries.

In conclusion, we see that in dealing with the concept of an infinite series of spatially infinite multiverses, we needed to develop a new kind of language that combined elements from the realms of philosophy and mathematics. With this new language, we were able to finally view the big cosmic picture and appreciate a few of its implications. The insights gained will, most likely, have little affect on one's daily life. However, it is somewhat comforting to know that each of us has an infinite genetic extension of ourselves that persists throughout infinite space and infinite time! In a sense, this is a kind of immortality granted to us by an also infinite cosmos. Because the laws of physics and chemistry remain constant from multiverse to multiverse, that means that each lifeform's unique genetic blueprint is never really eliminated from the cosmos. Rather, it is repeated infinitely within each multiverse and reappears from multiverse to multiverse. It has existed for all of past eternity, exists now, and will exist forever on into the future eternity.

Chapter 14

Conquest of Lotto

THIS FINAL CHAPTER IS YET another one that I had to give much consideration to before deciding to include in this volume. On the one hand, it has little to do with the book's general theme of paranormal phenomena and the ultimate nature of the world and cosmos we inhabit. Yet, on the other hand, it does, as the previous chapter did, deal with large numbers and the mathematics of probability theory. So, since I wrote it, loathe wasting good material, and it might be of some interest to some readers, I have decided to include it here. Those adverse to the notion gambling can either skip it or can just consider it as a short dissertation on the probabilistic nature of reality.

Here in my state of New Jersey we have a popular state lottery game called "Pick 6 Lotto" which is held twice weekly on Monday and Thursday nights. The basic idea of the game is that 49 numbered balls are placed into a Plexiglas machine that mixes them thoroughly after which six of the numbered balls are allowed to escape and collect in a transparent chute attached to the front of the machine. The entire process takes place without the numbered balls ever being touched by human hands.

Players over the age of eighteen can place a one dollar wager on a particular six number combination that they have selected and which they feel may be extracted from the machine during the next drawing of numbers or they can simply have the combination automatically selected for them via a "Quick Pick" option when they place their wager. Wagers are placed with various vendors around the state who have a computer terminal called a "Green Machine" located on their business premises. One can place as many wagers on a night's drawing as one likes with this game. These wagers can all be placed on a single combination or they can each be placed on a different combination.

Most people, the author included, realize that state lotto games like this are constructed so that the probability of any single player winning the top prize or "jackpot" by correctly guessing all six numbered balls that will be selected during a particular drawing is virtually zero. However, it is not quite zero and, eventually, after several drawings wherein none of the millions of people placing wagers was successful in having selected the particular six numbered balls that were drawn, one or more *extremely* lucky individuals do eventually manage to "hit" the jackpot.

Millions or even tens of millions of dollars may be won and the news soon spreads across the state. The millions of losers have each contributed their money which is then carefully divided into halves. Half of the money collected for a particular drawing goes to the winners and the other half goes to help the state fund various educational and institutional projects. The games top winners then busy themselves figuring out what they will do with the huge financial windfall they are about to collect. The losers have only their dreams of what their lives will be like *if* they manage to win the *next* drawing. Happily, the game is structured so that one can still win some money if one manages to "catch" only five or four or even three of the numbers drawn from the machine, but these "lower tier" wins are not "life changing" wins like a jackpot win. Yes, lower tier winners will get some of their weekly lottery "investment" back, but, without the much coveted jackpot win, one will, even with occasional lower tier prizes, always be losing money over time if playing this game regularly.

The average player of any state lottery is, I believe, only dimly aware of what is his actual probability of winning that state lottery's jackpot. Perhaps he is aware that his chances are somewhere between zero and slim, but is still really not sure of exactly how low that is.

For example, is seems that more and more states are adopting the "Pick 6" format with a set of 49 numbered balls. From a set of 49 numbered balls it is possible to select a total of 13,983,816 different subsets of 6 numbered balls. Since the *order* in which the balls escape the lotto machine is unimportant, each subset of 6 balls can be referred to as a "combination".

If, however, the order in which the 6 balls were drawn from the machine was important, then each subset of 6 balls would be called a "permutation". It would then be possible to make over ten billion different subsets of them from the set of 49 balls. If a player had to correctly guess which of the ten plus billion permutations would be drawn from the machine on any particular drawing night, then it might take years before someone finally "hit" the jackpot! Fortunately, that is not the case. While the computations necessary to determine the total number of combinations or permutations that can be derived from a set of numbered lotto balls is not overly complex, I will not burden the reader with them here.

To put the actual probability of any lotto player hitting the jackpot in a 6 numbered balls out of 49 numbered balls or "6/49" state lotto game into perspective, let us assume that a person starts playing a *semi*-weekly state lotto game on his 18th birthday and places wagers on different combinations of numbered balls in every drawing until he reaches the age of 100 years. Let us further assume that for each drawing he wagers one hundred dollars and, thus, purchases tickets containing 100 different 6 numbered ball combinations.

One might assume that with such steadfast playing, our hypothetical player would surely win the jackpot sometime during the 82 years that he plays. Unfortunately, even with such devotion, our player still has little hope of being a jackpot winner! During his 82 years of play, he would have placed wagers on a total of 8,528 drawings and played a grand total of 852,800 combinations of 6 numbered balls.

Although this is certainly a huge number of combinations, it is still small compared to the total number of possible numbered ball combinations that can be extracted from the lotto machine. In fact, our hypothetical lotto player would have spent most of his lifetime (and almost a million dollars!) and only played about 6.1 % of the 13,983,816 possible 6 numbered ball combinations that can be derived from the full set of 49 numbered balls. Thus, in a lifetime of playing, our player would only have a 6.1 % probability of a jackpot win which is equivalent to saying that he also has a 93.9 % probability of *not* hitting the jackpot. Obviously, if our player chooses to wager less than one hundred dollars per drawing and is not in every drawing, then his probability of a jackpot win during the 82 years of this trial period will be further reduced.

If, however, our hypothetical player did manage to wager one hundred dollars in every drawing, he could only "expect" to hit the jackpot, *on average*, once every 1344 years. This, however, would not be guaranteed. If, by some miracle, our player had acquired physical immortality and his state could continue having semi-weekly drawings infinitely, he would find that there would be some periods of 1344 years during which he won no jackpot at all and other periods in which he won two or even more jackpots! As tens of millennia slowly passed, he would eventually find that his jackpot wins averaged out to almost exactly one jackpot win per 1344 year period.

Of course, in spite of the above rather grim probability analysis, there is still the possibility that the very next wager a lotto player makes could hit the jackpot because it is not possible to predict exactly which drawing in a 1344 year period will be the one that produces one of the combinations that the player has a wager on. It is this extremely faint ray of hope that keeps most "regular" players in the game. And there are also the rather regular and the heavily advertised "rollovers" which can produce jackpots in excess of a hundred million dollars. Such rollovers occur when no one wins the jackpot for a particular drawing. That unclaimed money is then added to the next scheduled jackpot.

If one merely plays randomly selected combinations of numbered balls, then the only real way of very slightly increasing one's chance of hitting the jackpot is to play more different combinations in more drawings and those winners that win with such random number combinations are the product of sheer luck and nothing else.

A few years ago, a group of friends of mine decided to start a lottery "pool" and wanted me to join in. They hoped that by buying many tickets in every drawing that they could, in fact, force a jackpot win. Of course, that jackpot would have to be split among all the members of the pool each of which had to make a weekly financial contribution to remain a member in good standing. I performed some computations similar to the above and decided not to join. They proceeded without me and over the course of two years invested close to one hundred dollars every week. During the two year period that their pool existed, they did manage to win a few lower tier prizes, but it was only a small fraction of what they had "invested". Eventually, like most such pools, their members began to move or change jobs and the group fell apart.

One of the members of the disbanded lotto pool approached me one day to tell me that he wanted to continue to play the Pick 6 lotto game twice per week and wanted to know if I could figure out some system that would give him a significant chance of hitting the jackpot in his lifetime. I tried to explain that lotto combinations are random numbers and that one of the mathematical definitions of a random number is that it is a number that is *not* capable of being mathematically or scientifically predicted. Therefore, in theory, it should be impossible to develop any kind of system for any state lottery. My friend, however, did not want to hear this type of argument. He was convinced that such a system had to exist and, because he knew of my interest in mathematics and science, believed that I could help him find such a system.

Ordinarily, if it was not a friend making such a request, I would have forgotten completely about the matter and gotten on with my own affairs. But, in order to appease this person, I decided to see if there were any available "systems" that could actually dramatically increase one's chances of hitting a Pick 6 jackpot. I collected about a dozen such systems and decided to put them to a quick test. No, I did not actually put any money on any of them. Rather, I had another friend who was an expert in the Qbasic programming language write a simple general purpose lotto "test" program for me.

Thus, his computer program allowed me to input the criteria of a particular lotto system and then run thousands of simulated lotto drawings to see how many prizes (jackpot and low tier) the system was catching. The program then analyzed the frequency of wins in each prize category to see if it was higher than would be expected to occur if no system was used but, rather, just randomly selected combinations had been played. In other words, this computer program could tell me if the special combinations selected by the jackpot system I was testing were, in fact, actually winning more prizes than would have been won by playing an equal number of combinations picked by a random number combination generator like the Green Machine's "Quick Pick" feature.

As the reader can imagine, I was not surprised to discover that not one of the lotto systems I tested did any better or worse than random chance in terms of winning lotto prizes. My computer program could be set up to place wagers of the same amount on each drawing in a predetermined quantity of drawings and, when this was done, *every* system I tested lost money continuously over time. Alternatively, if one started an open ended run of drawings using an initial fixed amount of money to wager with on all of the drawings, it was only a matter of time before all of the money was lost. Even more discouraging was the fact that, although I personally ran tens of thousands of simulated drawings with the program, at no time did any of the lotto systems tested produce a single jackpot!

After a month or so of this analysis, I reported the results to my friend whose inquiry had started my investigation. He, however, was unimpressed by the negative results and continued to suggest that I could somehow come up with a magic system for him. He wanted something that was easy to use, inexpensive, and which would be producing lower tier prizes frequently enough to convince him that he was playing as "close" to the jackpot as possible. I decided to humor him and said I would take one more look at the problem, but if I could not develop something in a few days, then he would have to bother someone else for help because I had already spent way more of my time and effort on this matter than I had anticipated would be necessary.

Incredible as it may sound, I did manage to come up with a system for playing any Pick 6 lotto game which I found met all of my friend's requirements and it is a system with which he is satisfied. I have decided at this time to reveal this system so that others who are looking for something unusual to try will be able to use it. However, let me state immediately that it carries no "win guarantees" of any kind and that anyone who chooses to try it will be doing so at their own financial risk. For these reasons, any reader choosing to try my system should first do it strictly on paper for a few weeks to see if he is sufficiently impressed enough with its performance to actually want to wager any money on it.

To begin with, let me say that this system consists of two components. The first is a simple way of performing "field reduction" so that you can reduce the total quantity of numbered lotto balls being played from 49 balls down to only 36 balls from which you will have a slightly better than random chance of having all of the next drawing's 6 numbered balls selected. You will, therefore, always be using only 36 of the 49 numbers to make up the combinations on which you will wager. The second component of this system consists of a carefully constructed set of only 12 special combinations of the numbers derived by the first component.

So, once you have selected the 36 numbers to use, you will then arrange them into only 12 special six number combinations that you will play. Thus, this system will, *if* a single player decides to wager money on it, require him to spent exactly $12 per drawing. If a state has two lotto drawings per week and he wants to place bets in each drawing, then the cost will be $24 per week or $1248 per year.

This system is most affordable for the individual player when it is used by a small "pool" of players. For example, six people could comprise the pool and each would contribute $4 per week to play in the two weekly drawings. Any money won would then be divided equally amongst the members of the pool with each receiving 1/6 of the money won. They could also opt to just play one of the drawings per week and the cost to each pool member would only be $2 with all winnings still being shared equally amongst the members. They might take turns being the member who will select and then purchase the combinations for a particular week's drawing(s).

In order to decide which 36 of your state's 49 numbered lotto balls to use to make the 12 combinations which will be wagered on, it is very important that one know what the *last* two 6 numbered ball combinations drawn were. If one is serious about using this or any system to place wagers, it is a good idea to maintain an up to date running list of past drawn lotto combinations with the day and date of the drawing written in a column next to the actual combination drawn.

It is very important to know the last two 6 number lotto combinations drawn because there is a slightly increased probability that *none* of the numbers drawn in the *next* drawing will have been selected in the *prior* two drawings. You can, therefore, eliminate the numbers drawn from the last two drawings from the field of 49 numbers to perform this field reduction. If all of the numbers that make up the last two combinations are different, then that means you can eliminate those 12 numbers from the field of 49 and, thereby, reduce it to only 37 numbers. You must still find one more number to eliminate to get the total that will be used down to only 36 numbers. At this point, the list of combinations you have been keeping will come in very handy. Simply look back another 5 drawings past the last two drawn and find which one number appeared the *most*. Eliminate this number along with the last 12 selected from the prior two drawings so that 13 numbers have been eliminated from the total of 49 and reduced the number field to only 36.

In the event that one number *repeated* itself in the last two drawings, then you will only have 11 *different* numbers from these two drawings to eliminate from the total of 49. Once again, you will need to look back an additional 5 drawings past the last two to now find the *two* numbers which appeared the most. Eliminate these numbers along with the last 11 selected in the *prior* two drawings so that you have eliminated 13 numbered balls from the 49 and gotten the total quantity of numbered balls you will play down to only 36. In the unlikely event that more that two numbered balls repeated themselves in the last two drawings, then you will have to continue this process so that you always wind up with only 36 numbered balls to play for the *next* drawing. (Note: in the event that your state uses less that 49 numbered balls in its state lottery, then you might only have to eliminate the six numbered balls drawn in the last drawing plus a few more that appeared frequently in the 5 drawings prior to the last drawing.)

Let's now assume that one has begun using the above method for a state's 6/49 lotto game. Now what?

Begin by writing the numbers 1 through 49 in two vertical columns on a sheet of ruled notebook paper. These represent the 49 numbered balls used in the game. You can put numbered balls 1 through 25 in the left side column and numbered balls 26 through 49 in the right side column. Now go through these two columns of numbered balls and cross out the 13 that you will be *eliminating* from the number field of 49 numbered balls. When this is done, perform a quick check by counting the quantity of numbered balls that still remain in the two columns and make sure that there are exactly 36 of them. Once this is done, start with the lowest of the remaining numbered balls (which could be ball number 1 *if* you did not eliminate it) and place a capital letter "A" next to it. Move on to the next highest remaining numbered ball and place a capital letter "B" next to it. Then place a capital letter "C" next to the next highest remaining number after this and so on down your column of remaining numbered balls.

At some point, you will have reached the 26th remaining numbered ball and placed a capital "Z" next to it. For the next highest remaining numbered ball, place a *lower* case letter "a" next to it, then a lower case "b" next to the next highest remaining numbered ball after that and so on. If you have done everything correctly, then you should be placing a lower case letter "j" next to the last and highest one of the 36 numbered balls that you will be using. The reason that you have just assigned letters to the 36 numbers that you will be using is because you will now be given the 12 special combinations that you will play in the next drawing as combinations of *letters*. You will then simply substitute the numbered balls that were paired with each of your letters in the two vertical columns of your ruled notebook paper sheet into these 12 combinations of letters to form the 12 combinations of numbered balls which you will then play in your state's *next* lotto drawing.

Here are the 12 special combinations into which you must substitute the numbers on the balls to which you previously assigned capital and lower case letters of the alphabet:

ACENPR
ACEZbd
BDFMOQ
BDFYac
GIKTVX
GIKfhj
HJLSUW
HJLegi
MOQZbd
NPRYac
SUWfhj
TVXegi

As the reader may notice, each of the 36 numbered balls that has been selected is used exactly twice in this set of combinations. When deriving these combinations, I broke the group of 36 numbered balls up into 6 subgroups each containing 6 of the selected numbered balls and arranged them into a hexagonal pattern that somewhat resembles a snowflake. For this reason, I refer to this set of combinations as a "Snowflake Wheel". The 6 subgroups of 6 numbered balls were then rearranged so as to form 2 interlocking subsets that then each contained 18 of the 36 selected numbered balls. Finally, each of these subsets containing 18 numbered balls provided 6 of the 12 total combinations given above.

From 36 numbered balls, mathematical calculations show that it is possible to form 1,947,792 different combinations of 6 numbered balls.

Thus, *if* all 6 of the numbered balls drawn by the state for a particular drawing happen to be among the 36 that one has selected to use in the Snowflake Wheel, then, with 12 combinations of these selected balls wagered on, one should have a 12 in 1,947,792 or 1 in 162,316 chance of having the jackpot combination on one of the 12 combinations he has played. Obviously, this is not too impressive a chance of hitting the jackpot.

However, because of the high symmetry of the Snowflake Wheel's combinations and the highly scrambled numbered ball combinations that are produced by a state's lottery machine, I have found that the lotto number combinations that are drawn seem to have a general tendency to correspond to the 12 particular combinations found in the Snowflake Wheel. Although this property can not guarantee any level of win from any particular drawing, it should, over the course of several drawings, increase the overall probability of a win at any level. Thus, this property should increase somewhat the chance of having any particular jackpot combination drawn be among the 12 special combinations that are played in each drawing with this system.

For example, when I tested this system for my friend, I discovered that if I caught *all* 6 of the numbered balls drawn in the 36 I had selected based on eliminating the numbered balls selected in the two prior drawings, then I would win a 4th prize (which happens when one "catches" 3 of the 6 numbered balls drawn in one of his played combinations) close to 100% of the time. In considering a series of 10 such drawings wherein all 6 numbered balls were caught in the 36 played, the Snowflake Wheel managed to win four 3rd prizes (4 of the 6 numbered balls drawn on one of the combinations played) which pays significantly more than a 4th prize wins!

Since the 12 combinations of the Snowflake Wheel cost $12 to play per drawing, winning a 4th prize is not too impressive. In my state this prize is only about $3 to $5 and only recovers a small part of one's wager. However, a 3rd prize in my state pays about $50 so if one paid $120 to play ten times and won 4 third prizes of $50 each, then he would win a total of $200 and realize a *profit* of $80!

Of course, all of these optimistic win projections assume that one has caught *all* 6 of the numbered balls drawn in one's selected group of 36 numbered balls. If this does not happen, then the situation changes for the worse. In testing the system, I found that if I only caught 5 of the numbered balls drawn in my group of 36, then I would only win a 4th prize about 25% of the time or about once in every 4 drawings. In considering a series of 10 drawings wherein only 5 numbered balls drawn were caught in the 36 numbered balls that I had selected, the Snowflake Wheel only yielded a single 3rd prize. And, sadly, if only 4 of the numbered balls drawn were caught in the 36 I had selected, even a 4th prize became a rarity which occurred maybe once in 10 such drawings.

Since it is only possible to form 1,947,792 combinations of numbered balls from a reduced field of 36 numbered balls while the full field of 49 numbered balls can form 13,983,816 different combinations, this means that there is only a 1,947,792 / 13,983,816 probability of catching all 6 of the numbered balls drawn in any drawing in a *randomly* selected set of 36 numbered balls. This is equivalent to saying that there is about a 13.93 % or 1 in 7 chance of having all of the 6 numbered balls drawn in any particular lotto drawing appearing in a randomly selected set of 36 numbered balls.

The only way to enhance this somewhat low probability of having all 6 of the numbered balls drawn appear in one's selected group of 36 numbered balls is to make sure that they are *not* randomly selected as is usually done by the state's computerized lotto ticket terminal.

One must, therefore, be *very* careful in his selection of the 36 numbered balls which he will put into the 12 combinations that form the Snowflake Wheel. It is for this reason that I strongly suggest that anyone trying this system practice with it for several drawings *without* actually purchasing any tickets to make sure that he is proficient at selecting the 36 numbered balls which he will use to form his 12 combinations. This predrawing numbered ball selection process is a *critical* part of the system. If a player attempts to utilize the Snowflake Wheel by merely inserting randomly selected numbers into it, then he might as well save his time and effort and let the state's lottery machine pick the numbered ball combinations for him. He can then expect to catch all 6 of the numbered balls drawn in his randomly selected group of 36 numbered balls about once per month if he plays in every one of that month's 8 or 9 drawings. He would have spent over $100 and, perhaps, only won a single 3rd prize which would not recover the cost of his wagers.

In my short testing of this system, it did produce a few second prizes (5 of the 6 drawn numbered balls in one combination of the 12 played), but I did not see a jackpot with it. However, I am confident that with further testing it would, indeed, have produced that much sought after jackpot combination.

One of the nice things about using the Snowflake Wheel is that, after a drawing has taken place and one has written down the 6 numbered balls that were selected, he will immediately know if he has caught all six of them in this played group of 36 numbered balls. He will know that this has occurred if, upon inspection, he sees that *none* of the 6 selected numbered balls appears in the 13 that he *excluded* from the field of 49 balls so as to form the reduced field of 36 numbered balls 12 combinations of which he played.

For this reason, the player using this system should, after he has formed the 12 combinations that he will play, make an additional list of the 13 numbered balls that were excluded from those 12 combinations. That list should be kept handy and ready for reference when he finally sees the 6 numbered balls that were selected. Random chance says that, on average, one will only have all 6 of the drawn numbered balls in his group of 36 about once every 7 drawings. But, I think that the careful selection of the 36 numbered balls that are used may increase this probability a bit. Perhaps one will notice that he catches all 6 drawn balls every 5 or 6 drawings. At those times the excitement will mount as he carefully goes through his 12 played combinations to see if any one of them contains all 6 of the drawn balls.

Finally, let me remind the reader that when one is trying to win a 6/49 lotto jackpot, one is really not gambling in the same sense as when one places wagers in a typical casino. The odds in casino games are carefully formulated so that the gambler plays with only a slight disadvantage with respect to the house (note that the popular card game "Blackjack" or "21" is probably the only exception to this general rule).

This means that as one gambles in a casino, he will lose a dollar slightly more frequently than he will win one back and the excitement of gambling will be prolonged as long as possible. In the short term, a gambler in a casino may win a lot of money and then, just as quickly, lose it all back to the house so that, if he continues to gamble nonstop, he will eventually lose all of his starting supply of money. This process is based on the immutable laws of probability theory and guarantees the house a continuous and reliable source of income. This reliable revenue stream has helped build the casino "gaming" industry across the United States and the world which, only recently due to the worldwide recession, is showing any signs of slowing. People like the excitement of the casino environment with its bright lights and sound effects. If one is *not* prone to compulsive and financially ruinous gambling behavior, then such activity can probably be justified as a unique form of expensive adult entertainment. However, one should only gamble with money he is *fully* prepared to lose because, most of the time, that is exactly what will happen to it!

In the case of a state lotto game, however, the odds are formulated in such a way as to put the player at an enormous disadvantage with respect to the state. If one were to start with, say, $1000 and wager it on lotto tickets using randomly selected number combinations and then use one's winnings to buy additional tickets, one would find out that he would lose *all* of his starting money in a matter of a *few* drawings! Should one be lucky enough to win a second prize in the first 1000 combinations wagered on, this would only extend one's gambling for a few more drawings after the initial win and, eventually, all of the second prize money would be lost. The gambler can only hope to overcome all of this with a "life changing" jackpot win, the odds of which place it far beyond those of winning ordinary casino gambling games. In fact, one who seeks a lotto jackpot is really in search of a statistical miracle.

Lotto jackpots, in the final analysis, will, for most of those who win, be produced solely by blind chance. However, if a player wants to believe that he is truly playing with the maximum chance of having a jackpot win, then it is absolutely essential that he regularly use some sort of odds reducing system while playing. Perhaps the system given in this chapter will provide that.

Epilog

As we enter the 21ˢᵗ century, it seems to me that this particular century will be one of unusually great change for humanity. Indeed, I sense that we are rapidly approaching a critical fork in the path of civilization's ongoing evolution.

One branch of that fork leads to a veritable "Golden Age" of peace, prosperity, and freedom for all of humanity during which it will experience more progress in a single century or two than it has in all of its thousands of years of recorded history. Quite unfortunately, the other branch in the fork leads to a "Dark Age" that could be prolonged for millennia or even spell the extinction of the human race.

The branch that leads to the Golden Age assumes that many of the physical effects displayed in paranormal as well as ufological phenomena are relentlessly pursued until they are completely understood and incorporated into the body of known and accepted physics. Those effects will then be used to produce a new technology that will revolutionize every aspect of life on Earth. Fields as disparate as agriculture, astrophysics, biology, communication, manufacturing, medicine, metallurgy, psychology, and transportation will all undergo drastic expansion and revision.

With these rapid changes, the life of the average citizen of planet Earth will also change. Gone will be such obsolete anachronisms as disease, hunger, poverty, and war. Freed of these things that rob humanity of its dignity, humans will finally be readied to colonize the habitable moons and planets of our own solar system and then head out into the depths of nearby space to, eventually, make peaceful contact with some of the local extraterrestrial races whose craft have been spotted in Earth's atmosphere for thousands of years already. Within a generation of such regular contact, all fear of our nonhuman galactic neighbors will disappear and humans will actually be vacationing on the planets of nearby star systems. Trade amongst the various extraterrestrial races in our local region of space is probably already well established and the people of Earth will be able to enjoy the benefits of such trade. Who knows what wondrous artwork, foods, inventions, and medicines our spacefaring neighbors would be willing to trade for the diversity of items that humanity produces.

All of this, of course, assumes that humanity can demonstrate to our generally peaceful galactic neighbors that we are not a danger to their existence. Trust takes time to build and, for this reason, I always caution that, whenever extraterrestrial craft and their crews manifest in our planet's atmosphere, the greatest restraint should be made *not* to capture, destroy, or otherwise hinder them. In dealing with such beings, the "Golden Rule" (certainly appropriate for a "Golden Age"!) should be applied of doing onto others, even *non* human others, as one would have them do onto oneself. In the extremely rare situation in which a disabled craft with injured crew members is located, every effort to render assistance should be made. These are times when much can be learned about our neighbors and efforts to help them will not be forgotten and can help to enhance future contact with them.

If this is not done, then I fear that humanity is truly headed for a Dark Age in which all of the gains in science and technology made in the last several thousands of years will be lost, perhaps forever. It will be a time when humanity will degenerate into a level of existence little above that of prehistoric times. That small percentage of humanity that survives will break up into regional bands of hunter gatherers and that will be in continual conflict with each other. The skills of reading, writing, and mathematics will be lost along with farming, manufacturing, and government. Gone will be such things as electrical power, telephone communication, radio, television, and the internet. Once great cities will crumble away without anyone caring or, even if they did, having the materials needed to reverse the decay. Life for the average person will, thereafter, not significantly change from one millennium to the next.

I'm sure that some reading the above will simply dismiss what I've written as just so much alarmist thinking. But, the fact remains that today's world has an enormous number of pressures that are constantly trying to push us toward a new Dark Age.

Currently, over a billion of Earth's seven billion or so human inhabitants live in squalid conditions in slums and even garbage dumps. They must struggle just to provide the minimum number of calories per day to stay alive. They have little if any access to medical care and their unsanitary living conditions are conducive to the outbreak of all sorts of bacterial or viral plagues. In any kind of global calamity, these people, over a billion of them, would be the first to perish.

There is much talk about how the major powers, principally the United States and the former Soviet Union, have worked to make the world safer through various "strategic arms limitation" treaties. Because of these one might suppose that any sort of future nuclear war is virtually impossible. But, the reality of the situation is that these nations still possess thousands of missile deliverable nuclear weapons that can be launched within minutes at any time of the day or night. There have already been several incidents which could have caused these weapons to be deployed and were only stopped in the last few minutes before the first launch took place.

These weapons contain enough megatonnage of nuclear energy to destroy the surface of the Earth a hundred times over. Aside from this, the fires that would ensue if even a small percentage of them were used could trigger a "nuclear winter" which would darken the skies to the point where our planet's surface temperature plunged to subzero levels and all plant life and the animal life dependent upon it perished. Only a small percentage of the human race could wait out such a disaster as it subsisted on what food stores it could find.

With the use of the new centrifuge technology for isolating and concentrating the fissionable isotopes of uranium and plutonium, the cost of producing nuclear weapons has steady decreased. It won't be long before a tactical and even strategic nuclear force is within the reach of even a small Third World nation's despot. With this proliferation of nuclear weapons taking place, it is only a matter of time before various terrorist groups obtain them and begin using them to blackmail the major powers for money or other concessions. If they do not comply, then there is the risk of one of these "loose nukes" being detonated in a major city such as New York, London, or Paris with all of the chaos and destruction that would cause.

I won't go on listing more of the various pressures that are pushing us toward a Dark Age except to state that most of these pressures stem from a common cause: the competition between desperate groups of people for the Earth's limited sources. Those that lose in this ongoing competition feel left out and abandoned by those who are lucky enough to secure them for their own use. This exclusion leads to envy and anger and, eventually, to a world fraught with tensions that inevitably lead to armed conflicts and are directly responsible for the fact that, presently, the majority of our planet's almost two hundred nations are continually in a state of war either internally or with their nearest neighbors.

The only realistic solution I see that will allow humanity to escape a Dark Age destiny is for it to develop the Golden Age science and technologies that have been presented in this volume. Only with these, combined with a new spirit of concern for one's fellow human beings and our beautiful planet, will escape be possible.

The Golden Age is very real, still attainable, and the time to begin working toward it is *now*.

About the Author

K ENNETH W. B EHRENDT HAS BEEN a lifelong student of phenomena in the fields of ufology and the paranormal. Although professionally trained as a chemist, he has been investigating and writing about the UFO phenomenon since the early 1980's. He has had several personal sightings of UFOs during his lifetime which have convinced him in the reality of these objects and considers their study to be of great importance to humanity. He currently resides in suburban New Jersey where he continues his researches in the areas of ufology, paranormal phenomena, and free energy physics.